1695
-5
W9-BKS-728

Research Notes in Mathematics

Main Editors
A. Jeffrey, University of Newcastle-upon-Tyne
R. G. Douglas, State University of New York at Stony Brook

Editorial Board
F. F. Bonsall, University of Edinburgh
H. Brezis, Université de Paris
G. Fichera, Università di Roma
R. P. Gilbert, University of Delaware
K. Kirchgässner, Universität Stuttgart
R. E. Meyer, University of Wisconsin-Madison
J. Nitsche, Universität Freiburg
L. E. Payne, Cornell University
G. F. Roach, University of Strathclyde
I. N. Stewart, University of Warwick
S. J. Taylor, University of Virginia

DISCARDED
JENKS LRC
GORDON COLLEGE

Submission of proposals for consideration
Suggestions for publication, in the form of outlines and representative samples, are invited by the editorial board for assessment. Intending authors should contact either the main editor or another member of the editorial board, citing the relevant AMS subject classifications. Refereeing is by members of the board and other mathematical authorities in the topic concerned, located throughout the world.

Preparation of accepted manuscripts
On acceptance of a proposal, the publisher will supply full instructions for the preparation of manuscripts in a form suitable for direct photolithographic reproduction. Specially printed grid sheets are provided and a contribution is offered by the publisher towards the cost of typing.

Illustrations should be prepared by the authors, ready for direct reproduction without further improvement. The use of hand-drawn symbols should be avoided wherever possible, in order to maintain maximum clarity of the text.

The publisher will be pleased to give any guidance necessary during the preparation of a typescript, and will be happy to answer any queries.

Important note
In order to avoid later retyping, intending authors are strongly urged not to begin final preparation of a typescript before receiving the publisher's guidelines and special paper. In this way it is hoped to preserve the uniform appearance of the series.

Titles in this series

1 Improperly posed boundary value problems
A Carasso and A P Stone

2 Lie algebras generated by finite dimensional ideals
I N Stewart

3 Bifurcation problems in nonlinear elasticity
R W Dickey

4 Partial differential equations in the complex domain
D L Colton

5 Quasilinear hyperbolic systems and waves
A Jeffrey

6 Solution of boundary value problems by the method of integral operators
D L Colton

7 Taylor expansions and catastrophes
T Poston and I N Stewart

8 Function theoretic methods in differential equations
R P Gilbert and R J Weinacht

9 Differential topology with a view to applications
D R J Chillingworth

10 Characteristic classes of foliations
H V Pittie

11 Stochastic integration and generalized martingales
A U Kussmaul

12 Zeta-functions: An introduction to algebraic geometry
A D Thomas

13 Explicit *a priori* inequalities with applications to boundary value problems
V G Sigillito

14 Nonlinear diffusion
W E Fitzgibbon III and H F Walker

15 Unsolved problems concerning lattice points
J Hammer

16 Edge-colourings of graphs
S Fiorini and R J Wilson

17 Nonlinear analysis and mechanics: Heriot-Watt Symposium Volume I
R J Knops

18 Actions of fine abelian groups
C Kosniowski

19 Closed graph theorems and webbed spaces
M De Wilde

20 Singular perturbation techniques applied to integro-differential equations
H Grabmüller

21 Retarded functional differential equations: A global point of view
S E A Mohammed

22 Multiparameter spectral theory in Hilbert space
B D Sleeman

24 Mathematical modelling techniques
R Aris

25 Singular points of smooth mappings
C G Gibson

26 Nonlinear evolution equations solvable by the spectral transform
F Calogero

27 Nonlinear analysis and mechanics: Heriot-Watt Symposium Volume II
R J Knops

28 Constructive functional analysis
D S Bridges

29 Elongational flows: Aspects of the behaviour of model elasticoviscous fluids
C J S Petrie

30 Nonlinear analysis and mechanics: Heriot-Watt Symposium Volume III
R J Knops

31 Fractional calculus and integral transforms of generalized functions
A C McBride

32 Complex manifold techniques in theoretical physics
D E Lerner and P D Sommers

33 Hilbert's third problem: scissors congruence
C-H Sah

34 Graph theory and combinatorics
R J Wilson

35 The Tricomi equation with applications to the theory of plane transonic flow
A R Manwell

36 Abstract differential equations
S D Zaidman

37 Advances in twistor theory
L P Hughston and R S Ward

38 Operator theory and functional analysis
I Erdelyi

39 Nonlinear analysis and mechanics: Heriot-Watt Symposium Volume IV
R J Knops

40 Singular systems of differential equations
S L Campbell

41 N-dimensional crystallography
R L E Schwarzenberger

42 Nonlinear partial differential equations in physical problems
D Graffi

43 Shifts and periodicity for right invertible operators
D Przeworska-Rolewicz

44 Rings with chain conditions
A W Chatters and C R Hajarnavis

45 Moduli, deformations and classifications of compact complex manifolds
D Sundararaman

46 Nonlinear problems of analysis in geometry and mechanics
M Atteia, D Bancel and I Gumowski

47 Algorithmic methods in optimal control
W A Gruver and E Sachs

48 Abstract Cauchy problems and functional differential equations
F Kappel and W Schappacher

49 Sequence spaces
W H Ruckle

50 Recent contributions to nonlinear partial differential equations
H Berestycki and H Brezis

51 Subnormal operators
J B Conway

52 Wave propagation in viscoelastic media
F Mainardi

53 Nonlinear partial differential equations and their applications: Collège de France Seminar. Volume I
H Brezis and J L Lions

54 Geometry of Coxeter groups
H Hiller

55 Cusps of Gauss mappings
T Banchoff, T Gaffney and C McCrory

56 An approach to algebraic K-theory
A J Berrick

57 Convex analysis and optimization
J-P Aubin and R B Vintner

58 Convex analysis with applications in the differentiation of convex functions
J R Giles

59 Weak and variational methods for moving boundary problems
C M Elliott and J R Ockendon

60 Nonlinear partial differential equations and their applications: Collège de France Seminar. Volume II
H Brezis and J L Lions

61 Singular systems of differential equations II
S L Campbell

62 Rates of convergence in the central limit theorem
Peter Hall

63 Solution of differential equations by means of one-parameter groups
J M Hill

64 Hankel operators on Hilbert space
S C Power

65 Schrödinger-type operators with continuous spectra
M S P Eastham and H Kalf

66 Recent applications of generalized inverses
S L Campbell

67 Riesz and Fredholm theory in Banach algebra
B A Barnes, G J Murphy, M R F Smyth and T T West

68 Evolution equations and their applications
F Kappel and W Schappacher

69 Generalized solutions of Hamilton-Jacobi equations
P L Lions

70 Nonlinear partial differential equations and their applications: Collège de France Seminar. Volume III
H Brezis and J L Lions

71 Spectral theory and wave operators for the Schrödinger equation
A M Berthier

72 Approximation of Hilbert space operators I
D A Herrero

73 Vector valued Nevanlinna Theory
H J W Ziegler

74 Instability, nonexistence and weighted energy methods in fluid dynamics and related theories
B Straughan

75 Local bifurcation and symmetry
A Vanderbauwhede

76 Clifford analysis
F Brackx, R Delanghe and F Sommen

77 Nonlinear equivalence, reduction of PDEs to ODEs and fast convergent numerical methods
E E Rosinger

78 Free boundary problems, theory and applications. Volume I
A Fasano and M Primicerio

79 Free boundary problems, theory and applications. Volume II
A Fasano and M Primicerio

80 Symplectic geometry
A Crumeyrolle and J Grifone

81 An algorithmic analysis of a communication model with retransmission of flawed messages
D M Lucantoni

82 Geometric games and their applications
W H Ruckle

83 Additive groups of rings
S Feigelstock

84 Nonlinear partial differential equations and their applications: Collège de France Seminar. Volume IV
H Brezis and J L Lions

85 Multiplicative functionals on topological algebras
T Husain

86 Hamilton-Jacobi equations in Hilbert spaces
V Barbu and G Da Prato

87 Harmonic maps with symmetry, harmonic morphisms and deformations of metrics
P Baird

88 Similarity solutions of nonlinear partial differential equations
L Dresner

89 Contributions to nonlinear partial differential equations
C Bardos, A Damlamian, J I Díaz and J Hernández

90 Banach and Hilbert spaces of vector-valued functions
J Burbea and P Masani

91 Control and observation of neutral systems
D Salamon

92 Banach bundles, Banach modules and automorphisms of C*-algebras
M J Dupré and R M Gillette

93 Nonlinear partial differential equations and their applications: Collège de France Seminar. Volume V
H Brezis and J L Lions

94 Computer algebra in applied mathematics: an introduction to MACSYMA
R H Rand

95 Advances in nonlinear waves. Volume I
L Debnath

96 FC-groups
M J Tomkinson

97 Topics in relaxation and ellipsoidal methods
M Akgül

98 Analogue of the group algebra for topological semigroups
H Dzinotyiweyi

99 Stochastic functional differential equations
S E A Mohammed

100 Optimal control of variational inequalities
V Barbu

101 Partial differential equations and dynamical systems
W E Fitzgibbon III

102 Approximation of Hilbert space operators. Volume II
C Apostol, L A Fialkow, D A Herrero and D Voiculescu

103 Nondiscrete induction and iterative processes
V Ptak and F-A Potra

104 Analytic functions – growth aspects
O P Juneja and G P Kapoor

105 Theory of Tikhonov regularization for Fredholm equations of the first kind
C W Groetsch

106 Nonlinear partial differential equations and free boundaries
J I Díaz
107 Tight and taut immersions of manifolds
T E Cecil and P J Ryan
108 A layering method for viscous, incompressible L_p flows occupying R^n
A Douglis and E B Fabes
109 Nonlinear partial differential equations and their applications: Collège de France Seminar. Volume VI
H Brezis and J L Lions
110 Finite generalized quadrangles
S E Payne and J A Thas
111 Advances in nonlinear waves. Volume II
L Debnath
112 Topics in several complex variables
E Ramírez de Arellano and D Sundararaman
113 Differential equations, flow invariance and applications
N H Pavel
114 Geometrical combinatorics
F C Holroyd and R J Wilson

Geometrical combinatorics

F C Holroyd & R J Wilson (Editors)
The Open University, Milton Keynes

Geometrical combinatorics

WINN LIBRARY
Gordon College
Wenham, Mass. 01984

π

Pitman Advanced Publishing Program
BOSTON · LONDON · MELBOURNE

PITMAN PUBLISHING LIMITED
128 Long Acre, London WC2E 9AN

PITMAN PUBLISHING INC
1020 Plain Street, Marshfield, Massachusetts 02050

Associated Companies
Pitman Publishing Pty Ltd, Melbourne
Pitman Publishing New Zealand Ltd, Wellington
Copp Clark Pitman, Toronto

© F C Holroyd and R J Wilson 1984

First published 1984

AMS Subject Classifications: 05-06, 51-06

ISSN 0743-0337

Library of Congress Cataloging in Publication Data
Main entry under title:

Geometrical combinatorics.

Papers presented at a conference held at the Open University, Milton Keynes, U.K., on 9 Mar. 1984.
Bibliography: p.
1. Combinatorial geometry—Addresses, essays, lectures. 2. Combinatorial designs and configurations—Addresses, essays, lectures. I. Holroyd, F. C. (Fred C.) II. Wilson, Robin J.
QA167.G46 1984 516'.13 84-16576
ISBN 0-273-08675-8

QA
167
.G46
1984

British Library Cataloguing in Publication Data

Geometrical combinatorics.—(Research notes in mathematics, ISSN 0743-0337; 114)
1. Combinatorial geometry
I. Holroyd, F.C. II. Wilson, Robin J.
III. Series
516'.13 QA167

ISBN 0-273-08675-8

All rights reserved. No part of this publication may be reproduced, stored in a retrieval system, or transmitted, in any form or by any means, electronic, mechanical, photocopying, recording and/or otherwise, without the prior written permission of the publishers. This book may not be lent, resold, hired out or otherwise disposed of by way of trade in any form of binding or cover other than that in which it is published, without the prior consent of the publishers.

Reproduced and printed by photolithography
in Great Britain by Biddles Ltd, Guildford

Contents

† Speaker at the Conference

Preface

This book presents seven of the eight talks given at a one-day conference in Geometrical Combinatorics held at the Open University, Milton Keynes, U.K., on 9 March 1984. There is also an elementary proof, due to John Mason, of a result in the talk given by Norman Biggs.

How does one define "Geometrical Combinatorics"? If one is a formalist, one notes that it concerns itself in the main with incidence structures (finite or infinite), which can be described by incidence matrices; so perhaps one says : "It is the study of finite or infinite matrices of 0s and 1s." In our experience, however, a more creative approach is to organise a conference on the subject and see what transpires.

In addition to the wide range of topics presented in these talks, there is a variety of styles and approaches, from sharply focussed presentations of particular problems and their solutions, through expositions of present "states of the art", to Peter Cameron's exciting challenge to develop an infinite analogue of finite geometry.

We should like to thank Ian Dey, the Pro-Vice-Chancellor (Planning) of the Open University, for welcoming the participants and Roy Nelson and Howard Thomas for introducing the talks. Most of all we should like to thank Carole Fulcher for her excellent typing of the manuscript, and Bridget Buckley and the staff of Pitman Publishing for all their help.

The Open University
June 1984

Fred Holroyd
Robin J Wilson

N L BIGGS AND JOHN SHAWE-TAYLOR

1 Rotations and graphs with large girth

1. A THEOREM ABOUT ROTATIONS

Consider a triad of mutually orthogonal axes in 3-dimensional Euclidean space. The axes are composed of 6 unit vectors meeting at the origin and can be viewed as an unordered triple of unordered pairs of opposite vertices on the unit sphere. As they are unlabelled a rotation of $\pi/2$ about any one of the axes takes the triad to itself. We define a move on such a triad to be a rotation of $\pi/4$ about the current location of one of its axes. We call a sequence of moves simple if no move of the sequence is the same as the previous move. We now state the main result of this paper.

Theorem Given a simple sequence of moves $M_1, M_2, \ldots, M_k$, the only simple sequence which will regain initial position is the reverse sequence of moves $M_k, M_{k-1}, \ldots, M_2, M_1$.

This theorem may be viewed as a theorem about graphs if we consider the graph obtained by taking the triad positions as vertices and the moves as edges joining two positions. Clearly this graph is cubic and the theorem states that its components are infinite cubic trees. As there are uncountably many triads of axes, the graph will certainly have uncountably many components.

In order to prove the theorem we turn to the algebraic theory of sextets and sextet graphs. In Section 2 we introduce sextets and sextet graphs over a field with eighth roots of unity. In Section 3 the sextet graphs over finite fields as introduced by Biggs and Hoare [1] are considered. Results concerning the girth of these graphs are covered in Section 4. In Section 5 we state the correspondence between sextets and the triads of axes introduced above. This relation enables us to prove the theorem.

2. ALGEBRAIC THEORY OF SEXTETS

Let F be a field containing eighth roots of unity. The projective line over F, PG(1,F), may be considered as the set $L = F \cup \{\infty\}$, with the usual conventions about ∞. We shall call an unordered pair of distinct points {a,b} on L a duet, while a quartet is an unordered pair of duets whose cross-ratio is −1. Thus {{a,b},{c,d}}, or just {ab|cd} for short, is a quartet if and only if

$$\frac{(a-c)(b-d)}{(a-d)(b-c)} = -1.$$

A sextet is an unordered triple of duets, denoted by {ab|cd|ef}, such that {ab|cd}, {cd|ef} and {ef|ab} are all quartets.

Let σ be a primitive eighth root of unity in F and $i = \sigma^2$. Hence $i^2 = \sigma^4 = -1$; then $\{0\ \infty|1\ -1|i\ -i\}$ is an example of a sextet of L.

Sextets have appeared in the works of classical geometers, among others, Enriques and Edge [5]. They are sometimes referred to as 'regular sextuples'.

The group PGL(2,F) of projective linear transformations of the form

$$u \longrightarrow \frac{au+b}{cu+d} \quad (a,b,c,d \in F,\ ad-bc \neq 0)$$

can be considered to act on L with the usual conventions about ∞. This group acts sharply 3-transitively on L and maps sextets to sextets as it preserves the cross-ratio.

It is not hard to see that the points $0,\infty,1$ uniquely determine a sextet $\{0\ \infty|1\ -1|i\ -i\}$. Thus given any sextet $\{a_1\ a_2|a_3\ a_4|a_5\ a_6\}$ we can, by choosing the unique $g \in PGL(2,F)$ which maps a_1,a_2,a_3 to $0,\infty,1$ respectively, make g map the second sextet to the first. Hence PGL(2,F) acts transitively on sextets.

Proposition Let Q be a quartet over the field F containing a primitive eighth root of unity σ, and $T = \{a_1\ b_1|a_2\ b_2\}$ one of the three possible pairings of the elements of Q, so not necessarily a quartet. Then T determines an involution j_T of PGL(2,F) which maps a_i to b_i, i=1,2, and this involution has two distinct fixed points.

Proof Since PGL(2,F) preserves cross-ratio and acts transitively on quartets it is sufficient to consider the case when $Q = \{0\ \infty|1\ -1\}$. The three possible pairings of Q are Q itself, $R = \{0\ 1|\infty\ -1\}$ and $S = \{0\ -1|\infty\ 1\}$. The involution j_Q is $u \longrightarrow -1/u$. Its fixed points satisfy $u^2 + 1 = 0$ and so are i and −i. The involution j_R is $u \longrightarrow (1-u)/(1+u)$, so that its fixed points satisfy $u^2+2u-1 = 0$ or $u = -1\pm\sqrt{2}$. The involution j_S is $u \longrightarrow (u+1)/(u-1)$ with fixed points $u = 1\pm\sqrt{2}$. Since $\sqrt{2} = \sigma + \sigma^{-1}$, the two distinct fixed points of j_Q, j_R and j_S all lie in F.∎

We are now in a position to define the sextet graph, $S(F)$, over F. The vertices of $S(F)$ are the sextets of $PG(1,F)$. We define two sextets,

$$\{a_1\ a_2|b_1\ b_2|c_1\ c_2\} \text{ and } \{a_1\ a_2|b_1'\ b_2'|c_1'\ c_2'\},$$

to be adjacent in $S(F)$ if b_1', b_2' are the fixed points of j_R and c_1', c_2' are the fixed points of j_S, where

$$R = \{b_1\ c_1|b_2\ c_2\} \text{ and } S = \{b_1\ c_2|b_2\ c_1\}.$$

For example $\{0\ \infty|1\ -1|i\ -i\}$ is adjacent to $\{0\ \infty|\sigma\ -\sigma|\sigma^3\ -\sigma^3\}$ as $\sigma, -\sigma$ are the fixed points of the involution taking $1,-i$ to $i,-1$, respectively, while $\sigma^3, -\sigma^3$ are the fixed points of the involution taking $1,i$ to $-i,-1$, and $\{0\ \infty|\sigma\ -\sigma|\sigma^3\ -\sigma^3\}$ is indeed a sextet.

<u>Proposition</u> If $g \in PGL(2,F)$ and a_1, a_2 are the fixed points of j_R then ga_1, ga_2 are the fixed points of $gj_Rg^{-1} = j_{gR}$. Hence $PGL(2,F)$ acts as a group of automorphisms on the graph $S(F)$. ∎

In fact we shall see in Section 4 that the action of $PGL(2,F)$ is faithful on each component of $S(F)$. By mapping $0,\infty,1$ to $1,-1,i$, respectively, and $0,\infty,1$ to $i,-i,1$, respectively, we obtain two more distinct neighbours of $\{0\ \infty|1\ -1|i\ -i\}$ as the two images of $\{0\ \infty|\sigma\ -\sigma|\sigma^3\ -\sigma^3\}$. Since $PGL(2,F)$ acts transitively on sextets the graph is cubic, and by mapping $0,\infty,1$ to $0,\infty,\sigma$ we see that the definition of adjacency is in fact symmetric.

3. SEXTET GRAPHS OVER FINITE FIELDS

In this section we state briefly the results of Biggs and Hoare [1]. They consider the case of a sextet graph $S(F)$ where F is a finite field $GF(q)$, $g = p^n$. In this case, components of $S(F)$ are isomorphic for all powers of p for which $S(GF(p^n))$ exists. This leads to the definition of the sextet graph $S(p)$ of a prime p:

$$S(p) := \begin{cases} S_0(GF(p)), & p \equiv 1 \pmod 8 \\ S_0(GF(p^2)), & p \equiv 3,5,7 \pmod 8, \end{cases}$$

where $S_0(F)$ is the component of $S(F)$ containing the sextet $\{0\ \infty|1\ -1|i\ -i\}$. Note that the square of p is necessary in the cases $p \equiv 3,5,7 \pmod 8$ in order to guarantee eighth roots of unity in the field.

The order of $S(p)$ for various classes of p (mod 16) and a note as to whether the graph is bipartite or not is presented in Table 1.

p (mod 16)	order of S(p)	bipartite?
1	$p(p^2-1)/48$	no
3,5,11,13	$p^2(p^4-1)/24$	yes
7	$p(p^2-1)/24$	yes
9	$p(p^2-1)/24$	yes
15	$p(p^2-1)/48$	no

Table 1

As examples of these sextet graphs we consider three primes 7,3,17.

Example 1 (p=7) In this case S(7) is a component of the graph S(GF(49)), which has 4900 sextets and 350 components each of order $|S(7)| = 14$. The graph has girth 6 and is isomorphic to the graph of points and lines of PG(2,2). It is often called Heawood's graph and is in fact 4-arc transitive.

Example 2 (p=3) In this case S(3) is the only component of S(GF(9)), a graph with 30 vertices. The graph is 5-arc transitive and has girth 8. Its vertices can be represented by the partitions into pairs of 6 elements together with the pairs of the elements. Two vertices are adjacent if they represent a pair and a partition containing that pair (Tutte's 8-cage graph).

Example 3 (p=17) In this case S(17) is a component of S(GF(17)). The latter graph has 204 vertices and two components so that $|S(17)| = 102$. S(17) is 4-arc transitive and has girth 9. It is one of the four cubic graphs which are primitive and distance-transitive [2].

4. SYMMETRY AND GIRTH OF SEXTET GRAPHS

Consider the graph S(F) for some field F with an eighth root of unity σ. The maps

$$u \mapsto \frac{\sigma(u-1)}{u+1} \quad \text{and} \quad u \mapsto \frac{\sigma(u+1)}{u-1}$$

are both projective linear transformations and so automorphisms of S(F). We denote them by a and b respectively. The following sequence of vertices is obtained by taking $k_1 = \{0\ \infty | 1\ -1 | i\ -i\}$ and setting

$k_r = a^{r-1}(k_1) = b^{r-1}(k_1)$, $0 \leqq r \leqq 4$:

$$k_0 = \{1\ -1|-\sigma^3-i-\sigma\ \ \sigma^3-i+\sigma|\sigma^3+i+\sigma\ \ -\sigma^3+i-\sigma\}$$
$$k_1 = \{0\ \infty|1\ -1|i\ -i\}$$
$$k_2 = \{0\ \infty|\sigma\ -\sigma|\sigma^3\ -\sigma^3\}$$
$$k_3 = \{-\sigma^3-i+1\ \ -\sigma^3+i-1|\sigma\ -\sigma|\sigma^3+i-1\ \ \sigma^3-i+1\}$$
$$k_4 = \{\sigma^3+i+1\ \ (-\sigma^3-i-1)/3|-\sigma^3-i+1\ \ -\sigma^3+i-1|\sigma^3-i-1\ \ (-\sigma^3+i+1)/3\}.$$

As we saw in Section 2, k_1 and k_2 are adjacent and so k_0, k_1, k_2, k_3, k_4 is a 4-arc of S(F). The automorphism a shunts this 4-arc onto one of its successors, while b shunts it onto the other. Hence a and b generate a subgroup H of the automorphism group of S(F) which is 4-arc transitive. Note also that $H \subseteq PGL(2,F)$.

Suppose $e(\neq 1) \in PGL(2,F)$ fixes the 4-arc k_0, k_1, k_2, k_3, k_4. It must then stabilise the duets $\{1\ -1\}$, $\{0\ \infty\}$ and $\{\sigma\ -\sigma\}$ (and indeed $\{-\sigma^3-i+1\ \ -\sigma^3+i-1\}$ as well). If more than one pair is fixed pointwise then e is the identity. In the case when only $\{0\ \infty\}$ is fixed pointwise, -1 maps to 1, and e is the transformation $u \longrightarrow -u$, which does not fix k_4. So 0 and ∞ must be interchanged by e, and e has the form $u \longrightarrow b/u$, for some $b \in F$. In order for $\{1\ -1\}$ to be stabilised $b = 1$ or -1, while for $\{\sigma\ -\sigma\}$ to be stabilised $b = i$ or $-i$. Hence e cannot be chosen to fix the 4-arc k_0, k_1, k_2, k_3, k_4 and H acts regularly on the 4-arcs. This also implies that PGL(2,F) acts faithfully on $S_0(F)$, and so on each component of S(F); for if g is a non-trivial element which fixes some component S_1 and h maps S_1 to $S_0(F)$, then hgh^{-1} is a non-trivial element fixing $S_0(F)$.

We define a <u>word</u> of length ℓ in two non-commuting variables x and y to be a string of ℓ symbols $w = w_1 w_2 \ldots w_\ell$ where w_i is either x or y. If x and y are members of a monoid M then w(x,y) is also in M.

<u>Proposition</u> Let H be a group of automorphisms acting s-regularly on a graph G with H generated by two shunt automorphisms with respect to some s-arc. Then G has girth g if and only if the shortest word w such that w(a,b) is the identity in H has length g. ∎

This proposition allows us to investigate the girth of an s-regular graph by examining words in its generating automorphisms. Some values of g for various S(p) graphs are tabulated in Table 2. The third column of the table is the value of $c = (\log_2 n)/g$, where n is the number of vertices of S(p).

p	g	$c = (\log_2 n)/g$
3	8	0.613
7	6	0.634
17	9	0.741
31	15	0.618
73	22	0.635
193	25	0.688
233	28	0.678
313	30	0.676

Table 2

The significance of c stems from two known bounds on this parameter for cubic graphs [3]. The first is the trivial lower bound $c > \frac{1}{2}$ for all cubic graphs. The second is given by a general construction of a family of cubic graphs with large girth and gives c as approximately 1 for this family. For many years this was the best general construction for cubic graphs with large girth, but recently Weiss [6] has shown that the sextet graphs give a better value of c.

Theorem [6] Let n be the order and g the girth of the sextet graph S(p). If $p \equiv \pm 3, \pm 5, \pm 7 \pmod{16}$, then

$$\log_2 n \leqq 3g/4 + 3/2. \blacksquare$$

Hence Weiss's theorem shows that for infinitely many values of g there is a cubic graph with $c \leqq 3/4 + 3/(2g)$, a significant improvement of the value of 1 previously obtained. For our present purposes we need only the obvious corollary that, given any g_0, there is a prime p_0 such that the girth of $S(p_0)$ is at least g_0.

5. TRIADS OF AXES AGAIN

In this section we consider sextet graphs when the field of Section 2 is the complex field $\mathbb{C}$. In this case $L = PG(1, \mathbb{C}) = \mathbb{C} \cup \{\infty\}$, and the points of the projective line can be considered as points on the unit sphere in 3-dimensional Euclidean space. If the cross-ratio of two duets is -1, then the chords joining the two pairs are at an angle of $\pi/2$. So a triad of axes is certainly a sextet. However, not all sextets are triads of axes as the example $\{0\ \infty|2\ -2|2i\ -2i\}$ indicates. In order for a sextet to be a triad of axes its duets must be composed of antipodal points on the sphere; that is, each duet must be of the form $\{z, -1/\bar{z}\}$. We shall show that the component S_0 of $S(\mathbb{C})$ containing the sextet $\{0\ \infty|1\ -1|i\ -i\}$ is composed of sextets which are triads of axes, and that adjacency in S_0 corresponds to the rotations described in Section 1. Let θ be the complex number $e^{i\pi/4}$, and let a' and b' denote the automorphisms a and b of Section 4 with σ replaced by θ. Then we have

$$a'(u) = \frac{(1+\bar{\theta})}{(1+\bar{\theta})} a'(u) = \frac{(1+\theta)u - (1+\theta)}{(1+\bar{\theta})u + (1+\bar{\theta})},$$

and hence a' can be represented by the matrix

$$A = \begin{bmatrix} 1+\theta & -(1+\theta) \\ 1+\bar{\theta} & 1+\bar{\theta} \end{bmatrix}.$$

Similarly b' can be represented by the matrix

$$B = \begin{bmatrix} \bar{\theta}-i & \bar{\theta}-i \\ -(\theta+i) & \theta+i \end{bmatrix}.$$

Both of these matrices have the form

$$\begin{bmatrix} x & -y \\ \bar{y} & \bar{x} \end{bmatrix}$$

and so represent rotations (this result goes back to Cayley [4]). Since the composition of two rotations is a rotation, $H = \langle a', b' \rangle$ is a group of rotations. Clearly rotations map antipodal points to antipodal points and so preserve triads of axes. Since the group H acts transitively on the sextets in the component S_0 all these sextets are triads of axes.

Observe that the sextet k_2 of Section 4 with σ replaced by θ is a sextet of L and can be obtained from k_1 by a $\pi/4$ rotation about the axis $0,\infty$. Let this rotation be denoted by r. Since H acts transitively on the 1-arcs of S_0, we can find an element e of H mapping the edge (k_1, k_2) to any other edge (ℓ_1, ℓ_2) of S_0. Then $ere^{-1}(\ell_1) = \ell_2$. But ere^{-1} is a rotation of $\pi/4$ about the axis $e(\{0,\infty\})$, which is one of the duets of ℓ_1. We conclude that adjacency in S_0 corresponds to rotation of a triad about one of its axes by an angle of $\pi/4$.

Hence the theorem of Section 1 reduces to showing that $S_0(\mathbb{C})$ is an infinite cubic tree.

Suppose this is not the case. By the Proposition of Section 4 there is a word w of finite length g such that w(a',b') is the identity in H. For any ring R let $M_2(R)$ denote the ring of 2×2 matrices with elements in R. Consider the matrices

$$A(t) = \begin{bmatrix} t & -t \\ 1 & 1 \end{bmatrix}, \qquad B(t) = \begin{bmatrix} t & t \\ 1 & -1 \end{bmatrix},$$

which are elements of $M_2(\mathbb{Z}[t])$, where t is an indeterminate. Then $A(\theta)$ and $B(\theta)$ represent the projective linear transformations a' and b' respectively.

$$\text{Let } w(t) = w(A(t),B(t)) = \begin{bmatrix} \alpha_w(t) & \beta_w(t) \\ \gamma_w(t) & \delta_w(t) \end{bmatrix},$$

which is an element of $M_2(\mathbb{Z}[t])$, so that $\alpha_w(t)$, $\beta_w(t)$, $\gamma_w(t)$, $\delta_w(t)$ are polynomials in $\mathbb{Z}[t]$. If w(a',b') is the identity then $w(\theta)$ is a representation of the identity transformation. Hence $\alpha_w(\theta)-\delta_w(\theta) = 0$, $\beta_w(\theta) = 0$, $\gamma_w(\theta) = 0$. This means that t^4+1 divides each of these polynomials:

$$\begin{aligned} \alpha_w(t) - \delta_w(t) &= (t^4+1)q(t), \\ \beta_w(t) &= (t^4+1)r(t), \\ \gamma_w(t) &= (t^4+1)s(t), \end{aligned}$$

for some $q(t), r(t), s(t) \in \mathbb{Z}[t]$. For any $q(t) \in \mathbb{Z}[t]$, let $\tilde{q}(t)$ denote the polynomial q(t) with coefficients taken modulo p, for some fixed prime p. Then

$$\begin{aligned} \tilde{\alpha}_w(t) - \tilde{\delta}_w(t) &= (t^4+1)\tilde{q}(t), \\ \tilde{\beta}_w(t) &= (t^4+1)\tilde{r}(t), \\ \tilde{\gamma}_w(t) &= (t^4+1)\tilde{s}(t). \end{aligned}$$

Let F be a field of characteristic p with σ a primitive eighth root of unity in F, and consider the graph S(F). Let a and b be the transformations of Section 4 over the field F. Note that the following maps are homomorphisms.

$$M_2(\mathbb{Z}[t]) \xrightarrow{\text{mod } p} M_2(\mathbb{Z}_p[t]) \xrightarrow{t=\sigma} M_2(F)$$

This means that the transformation w(a,b) is represented by the matrix

$$\tilde{w}(\sigma) = \begin{bmatrix} \tilde{\alpha}_w(\sigma) & \tilde{\beta}_w(\sigma) \\ \tilde{\gamma}_w(\sigma) & \tilde{\delta}_w(\sigma) \end{bmatrix},$$

as this is the product $w(A(\sigma),B(\sigma))$ in the ring $M_2(F)$. But substituting σ in our expressions we obtain

$$\tilde{\alpha}_w(\sigma) - \tilde{\delta}_w(\sigma) = 0,$$
$$\tilde{\beta}_w(\sigma) = 0,$$
$$\tilde{\gamma}_w(\sigma) = 0.$$

So the word $w(a,b)$ represents the identity in the field F. Hence for all primes p, S(p) has a cycle of length g. This contradicts the corollary to Weiss's theorem. We conclude that $S_0(\mathbb{C})$ is indeed an infinite cubic tree, and the theorem of Section 1 follows.

REFERENCES

1. N L Biggs and M J Hoare, The sextet construction for cubic graphs, Combinatorica 3 (1983), 153-165.
2. N L Biggs and D H Smith, On trivalent graphs, Bull. London Math. Soc. 3 (1971), 155-158; MR 44#3902.
3. B Bollobás, Extremal Graph Theory (London Math. Soc. Monograph No. 11), Academic Press, 1978; MR80a:5120.
4. A Cayley, On the correspondence of homographies and rotations, Math. Ann. 15 (1879), 238-240.
5. W L Edge, Conics and orthogonal projectivities in a finite plane, Canad. J. Math. 8 (1956), 362-382; MR18-227.
6. A Weiss, Girths of bipartite sextet graphs, Combinatorica (to appear).

N L Biggs and John Shawe-Taylor,

Royal Holloway College,

Egham,

Surrey, TW20 OEX

AN ELEMENTARY PROOF OF THE THEOREM OF SECTION 1

Consider the matrices

$$A_1 = \begin{bmatrix} 1 & 0 & 0 \\ 0 & a & a \\ 0 & -a & a \end{bmatrix},\ A_2 = \begin{bmatrix} a & 0 & a \\ 0 & 1 & 0 \\ -a & 0 & a \end{bmatrix},\ A_3 = \begin{bmatrix} a & a & 0 \\ -a & a & 0 \\ 0 & 0 & 1 \end{bmatrix}$$

Lemma 1 Let $k \geqq 2$, and let $P = A_{r_1} A_{r_2} \ldots A_{r_k}$ where each r_s is 1,2 or 3 and $r_{s+1} \neq r_s (1 \leqq s < k)$. Then the entries $\{p_{ij}(a)\}$ of P are polynomials in a with integer coefficients, zero constant terms, leading coefficients equal to ±1 and degrees as follows.

(i) $p_{ij}(a)$ is of degree k-2 if $i = r_k$ and $j = r_1$;

(ii) $p_{ij}(a)$ is of degree k-1 if $i = r_k$ or $j = r_1$ but not both;

(iii) $p_{ij}(a)$ is of degree k otherwise.

Proof This follows easily by induction on k. ∎

Lemma 2 If $p_{ij}(a)$ is one of the polynomials described in Lemma 1, then neither of the equations $p_{ij}(a)+1 = 0$, $p_{ij}(a)-1 = 0$ has $1/\sqrt{2}$ as a root.

Proof By Gauss' lemma for polynomials (see [1, Chapter 3]), the polynomial $2a^2-1$ cannot be a factor of either $p_{ij}(a) + 1$ or $p_{ij}(a) - 1$. ∎

Now let the starting position of the triad of axes be the x,y and z axes of 3-dimensional Euclidean space, and let $M_1, \ldots, M_k$ be a simple sequence of moves, as described in Section 1. Then this sequence of moves is described by a matrix product $B_k B_{k-1} \ldots B_1$, where:

(i) B_1 is A_1, A_2 or A_3 with a set equal to $1/\sqrt{2}$ or $-1/\sqrt{2}$;

(ii) for each $i = 2, \ldots, k$, B_i is $(B_{i-1} \ldots B_1) A_j (B_{i-1} \ldots B_1)^{-1}$, where A_j is A_1, A_2 or A_3 with a set equal to $1/\sqrt{2}$ or $-1/\sqrt{2}$.

Since a move can be described equally well by a rotation or its inverse, we may set a equal to $1/\sqrt{2}$ each time. Then $B_k B_{k-1} \ldots B_1$ may be rewritten as

$A_{r_1}A_{r_2}\ldots A_{r_k}$ where each r_j is 1,2 or 3, no two consecutive r_j's being equal. In order for this to bring the triad to its starting position, $A_{r_1}\ldots A_{r_k}$ would have to contain six zero entries and three entries equal to ±1. But Lemmas 1 and 2 show that no entry can be equal to ±1. ∎

REFERENCE

1. I N Herstein, Topics in Algebra, Blaisdell, 1964; MR 30#2028.

J H Mason,
Faculty of Mathematics,
The Open University,
Walton Hall,
Milton Keynes, MK7 6AA.

PETER J CAMERON

2 Infinite versions of some topics in finite geometry

1. INTRODUCTION

The history of finite geometry is much shorter than that of the more classical branches of the subject; the oldest researches in finite geometry are not much more than two centuries old. But in its comparatively short lifetime, finite geometry has developed a style of its own, both in the structures it studies and in the techniques it uses.

Many of its concepts can have the assumption of finiteness removed, sometimes after a certain amount of redefinition. This leads to problems of "infinite" geometry which still have the distinctive flavour of finite geometry, and (more important) are often amenable to the finite techniques or modifications of them.

This article attempts to describe some of these infinite problems. Usually the state of knowledge is fairly primitive, and much of the article is speculation on what might be done, rather than description of results.

In the next three sections, we consider specific topics involving finite projective or linear spaces, where first the finiteness of order, then the finiteness of dimension, then both, are relaxed. The topics are orbit theorems for collineation groups (inequalities connecting the numbers of point and line orbits); a permutation-group-theoretic characterisation of projective and affine groups which appeared first in the model theory literature; and a characterisation of subsets of projective spaces by certain line-intersection properties, used in studying groups generated by transvection subgroups. The next two sections consider analogues of finite structures such as generalised polygons, partial geometries and Steiner systems, in which some of the parameters are allowed to be infinite.

2. FINITE DIMENSION, INFINITE ORDER

A familiar result in finite geometry asserts that, under suitable conditions, a group of automorphisms of an incidence structure has at least as many line orbits as point orbits. One class of structures for which this orbit theorem holds is that of <u>linear spaces</u>, in which any two points lie on a unique line (with the nondegeneracy conditions that there is at least one line, any line has at least two points on it, and no line contains every point). The proof goes like this: the incidence matrix A of such a structure is shown to have rank equal to the number of points; this matrix intertwines the permutation representations of the automorphism group on points and lines, whence the representation on lines contains that on points as a constituent; and the conclusion follows by character theory.

In 1968, Valette (see Buekenhout [1]) pointed out that the theorem is false in the infinite case. Consider the incidence structure whose point set is the closed unit disc D in $\mathbb{R}^2$, and whose lines are the non-trivial intersections of D with lines of $\mathbb{R}^2$. (This is the real hyperbolic plane with the points at infinity adjoined.) It is easily checked that the full automorphism group has two point orbits (the interior and the boundary points) but only one line orbit.

An interesting open problem (J Doyen, personal communication) is to construct line-transitive examples with more than two point-orbits.

Some positive results may perhaps exist, however. First let us re-formulate the finite proof. Let $V(\mathcal{P})$ and $V(\mathcal{L})$ be the vector spaces of real functions on the point set $\mathcal{P}$ and line set $\mathcal{L}$ respectively. Then the incidence matrix represents a linear transformation $\theta: V(\mathcal{P}) \to V(\mathcal{L})$, and the first part of the proof asserts that θ is injective. The intertwining property says that θ maps the space $V(\mathcal{P})^G$ of G-invariant functions in $V(\mathcal{P})$ into the corresponding space $V(\mathcal{L})^G$, where G is any group of automorphisms; so $\dim V(\mathcal{P})^G \leqq \dim V(\mathcal{L})^G$. But $\dim V(\mathcal{P})^G$ is just the number of G-orbits in $\mathcal{P}$, since the characteristic functions of the orbits form a basis for $V(\mathcal{P})^G$. A similar remark holds for $V(\mathcal{L})^G$, proving the orbit theorem.

The transformation θ is defined by the rule

$$(f\theta)(\ell) = \sum_{p\in\ell} f(p)$$

for $f \in V(\mathcal{P})$, $\ell\in\mathcal{L}$. In this form, there is some hope of generalising the result, at least to linear spaces with "nice" measures defined on the point and line sets. We define $\theta: L^1(\mathcal{P}) \to L^1(\mathcal{L})$ by

$$(f\theta)(\ell) = \int_\ell f$$

for $f \in L^1(\mathcal{P})$, $\ell\in\mathcal{L}$. The first requirement is then a Fubini-type theorem to guarantee that the integral exists for almost all ℓ and that $f\theta \in L^1(L)$. Then we would need to show that θ is injective. Finally, we would require that, for suitably 'nice' automorphism groups G, the characteristic functions of G-orbits are integrable. If all this could be done, we could conclude that the number of non-null G-orbits on points is at most the number of non-null line orbits, if these numbers are finite. Note that this conclusion does hold in Valette's example, since the boundary points form a null set.

A projective plane is simply a linear space whose dual is also a linear space. Thus a collineation group of a finite projective plane has equally many point and line orbits. As far as I am aware, no example is yet known of a projective plane with a collineation group having different numbers of point and line orbits.

An even more speculative approach to an infinite orbit theorem would be to consider the projective plane over a local field, and 'approximate' it by the Hjelmslev planes defined over the finite rings obtained as quotients of the ring of integers by powers of its maximal ideal.

3. INFINITE DIMENSION, FINITE ORDER

Recently a characterisation of projective and affine spaces of countable dimension over finite fields has appeared, somewhat surprisingly, in the model-theoretic literature. Zil'ber [8] obtained a number-theoretic proof of a result equivalent to the following theorem; Cherlin (see [4]) gave an alternative proof depending on the classification of finite simple groups. Later, Neumann [5] gave a proof of the theorem in the context of permutation groups. Let Ω be a countable set. A moiety of Ω is an infinite subset with infinite complement. Any group of permutations of Ω permutes the moieties among themselves.

Theorem (Zil'ber, Cherlin and Neumann). Let G be a primitive permutation group on a countable set Ω, and suppose that every orbit of G on moieties is uncountable. Then either

(i) G is k-fold transitive on Ω, for all k; or

(ii) Ω has the structure of a projective or affine space over a finite field, and G acts as a group of collineations of Ω. ∎

It would take us too far afield to explain why the model theorists required this result; but a sketch of the Cherlin-Neumann proof is instructive. First, it is shown that any finite subset Δ of Ω is contained in a finite subset Δ' with the property that the subgroup of G stabilising Δ' pointwise is transitive on its complement. Thus, G is a Jordan group, and the sets Δ' are the flats of a geometry (to be precise, a matroid, or geometric lattice) on Ω. Moreover, the intervals in this lattice are finite and admit finite Jordan groups. Such groups are doubly transitive, by Jordan's theorem, and hence known, using the classification of the finite simple groups. The proof is completed by investigating how these intervals can be fitted together.

Thus, the difficult part of the proof involves finite geometry and group theory. It would be nice to have a proof which does not use machinery quite as heavy as the classification of finite simple groups. We do not even need the full list of finite Jordan groups; knowledge of those of large dimension would be enough. I do not know whether Zil'ber's proof gives any information about this.

A related question which has interested model-theorists is due to Lachlan: is there an $\aleph_0$-categorical pseudoplane? (A pseudoplane is an incidence structure in which any point lies on infinitely many lines, and dually, and any two points lie on only finitely many lines, and dually. Finite geometers do not need to understand $\aleph_0$-categoricity; according to the Ryll-Nardzewski theorem, it is equivalent, for countable structures, to the condition that the automorphism group has only finitely many orbits on k-tuples for all k.) No examples are known. However, Wiegand [7] found examples satisfying 'half' of the conditions. His incidence structure consists of the points and irreducible curves in the affine plane over the algebraic closure of a finite field — curiously, the same structure is obtained, whichever finite field is used! Any curve contains infinitely many points, any two curves meet in just finitely many points, and the automorphism group acts k-fold transitively on points for every k.

4. INFINITE DIMENSION AND ORDER

In the 1960's, McLaughlin found all the (finite-dimensional) irreducible linear groups generated by full transvection subgroups. (A transvection is a map $t_{x,\phi}: v \mapsto v + \phi(v).x$ of a vector space V, where $x \in V$ and $\phi \in V^*$; the corresponding transvection subgroup is generated by all $t_{x,\phi}$ as x and ϕ range over 1-dimensional subspaces of V and V*.) In general, only the special linear groups and symplectic groups arise; but over the field GF(2), there are some other examples, namely orthogonal groups and symmetric groups.

Recently, Jonathan Hall and I found a way to re-do McLaughlin's work without making any assumption of finite dimensionality. Our conclusions are very similar to his, but there is not a unique "special linear" or "symplectic" group on an infinite-dimensional vector space. For example, there can exist inequivalent symplectic forms, and we have no nice classification of them.

One of the key parts of the argument is entirely geometric, and is of a type familiar in finite geometry, concerning sets of points in projective space having prescribed intersections with lines. For comparision, probably the simplest result of this type asserts that if a set S of points meets every line ℓ in 0,1 or all points of ℓ, then S is the point set of a projective subspace. Our result gives a precise but more complicated description under a weaker hypothesis.

Theorem Suppose that the points of a projective space $\mathcal{P}$ are coloured with two colours in such a way that any line ℓ meets each colour class in 0,1, all but one, or all points of ℓ. Suppose further that either the order of the projective space is greater than two, or no plane meets a colour class in a triangle. Then there is a chain $\mathcal{C}$ of subspaces of $\mathcal{P}$, and a map f from points of $\mathcal{P}$ to $\mathcal{C}$, such that:

(i) for any point p, $p \in A(p) := f(p) \setminus \bigcup\{S \mid S \in \mathcal{C}, S \subset f(p)\}$;

(ii) the set A(p) in (i) is contained in a single colour class. ∎

Remarks

1. In the finite-dimensional case, $\mathcal{C}$ is a finite chain $S_0 \subset S_1 \subset S_2 \subset \ldots$, and the assertion is that the sets $S_i \setminus S_{i-1}$ (i = 1,2,...) belong to alternate colour classes.

2. The second hypothesis is necessary, since over GF(2) the condition about line intersections is entirely vacuous. This gives some sort of geometric explanation for the exceptional transvection groups over GF(2).

Let me indicate how this theorem is used. Suppose that G is an irreducible collineation group generated by transvection subgroups. It is shown that an orbit of G on transvection centres satisfies the line intersection condition of the theorem. So, over fields other than GF(2) (or with an extra hypothesis in that case), we deduce the existence of a chain $\mathcal{C}$ as in the theorem. Let $t_{x,\phi}$ be any transvection in G, and p = <x> its centre. By definition $t_{x,\phi}$ fixes f(p) and all subspaces above it in C. Also, $t_{x,\phi}$ fixes $\cup \{S | S \in C, S \subset f(p)\} = W$, and W does not contain p; so $W \subseteq \ker \phi$, whence every subspace below f(p) in C is fixed. Thus all transvections fix all subspaces in C, and so does the group they generate, namely G. Irreducibility of G now forces the chain to be trivial, whence every point is a transvection centre, and G is transitive on points.

5. GENERALISED POLYGONS AND PARTIAL GEOMETRIES

A generalised quadrangle with parameters s,t is an incidence structure having s + 1 points per line, t + 1 lines per point, any two points on at most one line, and having the property that if the point p is not on the line ℓ then there is a unique point of ℓ collinear with p. The importance of these and related objects in finite geometry is well known.

While investigating infinite permutation groups having equally many orbits on triples and quadruples [3], I came across the following curious result: there does not exist a generalised quadrangle having s = 2 and t infinite. The proof contained implicitly the facts (well-known in the finite case) that any generalised quadrangle with s = 2 has t = 1, 2 or 4, and that the quadrangles in each case are unique.

I communicated this result to Bill Kantor, who showed that there is no quadrangle with s = 3 and t infinite. His proof was group-theoretic, and again gave much more information; he showed that s = 3 implies that $t \leq 9$, and he proved the uniqueness of the quadrangle with s = 3, t = 9.

The increasing difficulty of uniqueness proofs for generalised quadrangles may have contributed to the fact that the obvious generalisation of these results (to s finite, t infinite) has not yet been achieved. As a first step, one might try to re-do the cases s = 2 and s = 3 in a way which does not involve producing such detailed information about the finite objects.

Still further generalisations could be considered. For example, what about arbitrary generalised 2n-gons with s finite and t infinite? (All finite generalised hexagons with s = 2 are known, so this would be a reasonable starting point.)

However, further work on the permutation group problem which provided the original motivation suggests that it might be more useful to look at partial geometries instead. These are incidence structures with three parameters

s, t, α; they satisfy the same conditions as generalised quadrangles except that, if $p \notin \ell$, then exactly α points of ℓ are collinear with p. (So a generalised quadrangle is a partial geometry with $\alpha = 1$.)

It is easy to see that a partial geometry must satisfy $\alpha \leq s + 1$. Moreover, if $\alpha = s + 1$, then the geometry is a linear space; and if $\alpha = s$, it is a 'transversal design' (equivalent to $s - 1$ mutually orthogonal Latin squares of order t). In either case, for fixed s, examples exist with t infinite, or arbitrarily large finite. It is known, however, that if $\alpha < s$ and t is finite, then t is bounded by a function of s. The problem, then, is to decide whether there can exist examples with s finite, $\alpha < s$, and t infinite. The first case where the answer is not known is $s = 3$, $\alpha = 2$; this was also the case that arose in the permutation group problem.

6. STEINER SYSTEMS, PERFECT CODES

In contrast to the rarity of finite Steiner systems with large values of t (none are yet known with $t \geq 6$), infinite examples may be constructed freely. Given t and k (with $t < k$), start with a partial Steiner system (i.e. any t points lie in at most one block, and any block contains at most k points). Adjoin, alternately, new blocks incident with those t-tuples of points not already incident with a block, and new points so that each existing block has its full quota of k points. After countably many steps, we have a Steiner system.

Many other objects can be constructed freely in this way, (for example, projective planes) and we can even build in automorphisms; thus, for any t, there is a Steiner system with $k = t + 1$ having a t-fold transitive automorphism group. Just where does the dividing line occur between what is freely possible and what is not? For example, for Steiner systems with $t = 2$, Pasch's axiom forces the system to be a projective space; and as we've seen, infinite generalised quadrangles with $s = 2$ or $s = 3$ are impossible.

It is not hard to show that any infinite Steiner triple system ($t = 2, k = 3$) is extendable to a Steiner quadruple system ($t = 3$, $k = 4$); one of the big unsolved problems is whether this is true for finite Steiner systems (see Quackenbush [6] for discussion).

Steiner systems with $k = 2t - 1$ could occur as the supports of the words of minimum weight in perfect binary codes. Recently, Dave Cohen (personal communication) has found free constructions for perfect codes. Interesting connections await investigation here. In particular, the existence of linear perfect binary codes is known to be equivalent to that of Steiner systems satisfying various 'configuration theorems' such as the symmetric difference property and a modified form of Pasch's axiom (see [2]). Do these objects exist?

REFERENCES

1. F Buekenhout, Homogénéité des espaces linéaires et des systèmes de blocs, Math. Z. 104 (1968), 144-146.
2. P J Cameron, Characterisations of some Steiner systems, parallelisms and biplanes, Math. Z. 136 (1974), 31-39. MR 50#1908.
3. P J Cameron, Orbits of permutation groups on unordered sets, II, J. London Math. Soc. (2), 23 (1981), 249-264. MR 82k: 20005.
4. G Cherlin, L Harrington and A H Lachlan, $\aleph_0$-categorical $\aleph_0$-stable structures, to appear.
5. P M Neumann, Some primitive permutation groups, to appear.
6. R W Quackenbush, Algebraic speculations about Steiner systems, Ann. Discr. Math. 7 (1980), 25-35.
7. R Wiegand, Homeomorphisms of affine surfaces over a finite field, J. London Math. Soc. (2), 18 (1978), 28-32; MR 58#10933.
8. B I Zil'ber, Totally categorical theories: structural properties and the non-finite axiomatizability, Lecture Notes in Math. 834, Springer-Verlag, Berlin 1980, pp. 381-410; MR 82m: 03045.

Peter J Cameron,
Merton College,
Oxford.

J I HALL

3 Symplectic geometry and mapping class groups

We are concerned here with a problem initially about graphs and linear transformations, but quite soon we see it set more naturally in the context of symplectic geometry over the field with two elements. One attraction of the problem is its application in many and diverse mathematical situations. We shall discuss in detail how the problem arises in the realm of surface topology and mapping class groups of 2-manifolds.

1. THE PROBLEM

Let D be a connected graph with vertex set X, and let $\underline{X}$ be the power set of X thought of as a vector space over $\mathbb{F}_2$ with distinguished basis $\{\underline{x} \mid x \in X\}$. For each $x \in X$, the linear transformation $\tau(x)$ of $\underline{X}$ is given by the action

$$\tau(x) : \begin{cases} \underline{x} \longmapsto \underline{x} & \\ \underline{y} \longmapsto \underline{y}+\underline{x} & \text{when } x \text{ is adjacent to } y \text{ in D} \\ \underline{y} \longmapsto \underline{y} & \text{when } x \text{ is not adjacent to } y. \end{cases}$$

With D we associate the group of linear transformations of $\underline{X}$ generated by the various $\tau(x)$:

$$\overline{W}(D) = \langle \tau(x) \mid x \in X \rangle.$$

The fundamental question is:

<u>Problem</u> What is $\overline{W}(D)$?

Of course, as stated the problem is too vague, particularly so considering how exact is the recipe for constructing $\overline{W}(D)$. The type of answer given and the degree of precision required have varied with the desired application.

Before continuing we give the isomorphism types of some examples which are typical of the particular solution presented here and which also suggest the possibility of varied application.

$$\overline{W}\left(\underbrace{\circ\!\!-\!\!\circ\!\!-\!\!\circ\!\!-\!\!-\!\!-\!\!-\!\!\circ\!\!-\!\!\circ}_{m \text{ vertices}}\right) \cong S_{m+1}$$

$$\overline{W}\left(\underbrace{\circ\!\!-\!\!\circ\!\!-\!\!\circ\!\!-\!\!-\!\!-\overset{\circ}{\underset{\circ}{<}}}_{m \text{ vertices}}\right) \cong C_2^k \wr S_m$$

where $k = m-2$ for m even,
and $k = m-1$ for m odd

$$\overline{W}\left(\circ\!\!-\!\!\circ\!\!-\!\!\overset{\overset{\circ}{|}}{\circ}\!\!-\!\!\circ\!\!-\!\!\circ\right) \cong O_6^-(2)$$

$$\overline{W}\left(\circ\!\!-\!\!\circ\!\!-\!\!\overset{\overset{\circ}{|}}{\circ}\!\!-\!\!\circ\!\!-\!\!\circ\!\!-\!\!\circ\right) \cong Sp_6(2)$$

$$\overline{W}\left(\circ\!\!-\!\!\circ\!\!-\!\!\overset{\overset{\circ}{|}}{\circ}\!\!-\!\!\circ\!\!-\!\!\circ\!\!-\!\!\circ\!\!-\!\!\circ\right) \cong O_8^+(2)$$

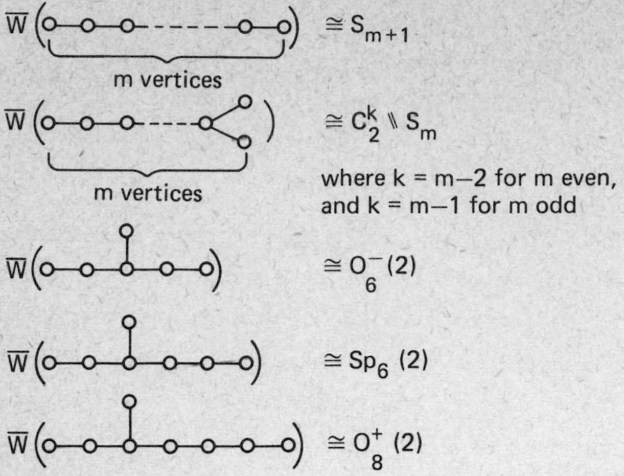

Here S_m is the symmetric group of degree m and $C_2^k \wr S_m$ is a split extension of an elementary abelian 2-group of order 2^k by S_m. As $\mathbb{F}_2 S_m$-module the subgroup C_2^k of the examples (k = m-2 or m-1 as m is even or odd) is the unique non-trivial irreducible consituent of the $\mathbb{F}_2$-permutation module for S_m, the "natural" irreducible $\mathbb{F}_2 S_m$-module. $Sp_{2m}(2)$ is the full group of isometries for a non-degenerate symplectic form on the space $\mathbb{F}_2^{2m}$. $O_{2m}^\varepsilon(2)$ is the full isometry group for a non-degenerate quadratic form on $\mathbb{F}_2^{2m}$; $\varepsilon = +$ or $-$ as a maximal totally singular subspace has dimension m or m-1. (See [2] for geometric details.)

In these examples we recognise the Coxeter or Dynkin diagrams of types A_m, D_m, E_6, E_7 and E_8 (see [9]). $\underline{X}$ is then the associated root lattice read modulo 2, so the groups $\overline{W}(D)$ are the usual Weyl groups factored when necessary by the subgroup $\{\pm I\}$ which acts trivially on $\underline{X}$. (This is our motivation for the $\overline{W}$ notation.)

The Weyl groups are integral Euclidean groups generated by orthogonal reflections. In the examples the linear transformations $\tau(x)$ are transvections, the mod 2 reductions of certain of the reflections. This suggests that some geometry may lie behind the structure of $\overline{W}(D)$, and this is indeed the case. The $\mathbb{F}_2$-adjacency matrix J of D is the Gram matrix with respect to the basis $\{\underline{x} \mid x \in X\}$ for a symplectic form f on $\underline{X}$ left invariant by $\overline{W}(D)$. That is, the (x,y)-entry of J is $f(x,y)$; and $f(\underline{u},\underline{v}) = f(\underline{u}\tau(x),\underline{v}\tau(x))$, for all $\underline{u},\underline{v} \in \underline{X}$ and all $x \in X$. (Indeed the function $q(\underline{x}) = 1$, $x \in X$, extends uniquely to a quadratic form on $\underline{X}$ associated with f and invariant under $\overline{W}(D)$.) The linear transformation $\tau(x)$, $x \in X$, is the symplectic transvection $\tau(x)$ of $\underline{X}$ with centre $\underline{x}$:

$$\tau(\underline{x}): \underline{y} \mapsto \underline{y} + f(\underline{y},\underline{x})\underline{x}.$$

We see therefore that the problem finds a natural setting within the classification of subgroups of symplectic groups which are generated by transvections. The following theorem gives one solution to the problem.

Theorem Assume D is connected and finite with more than one vertex. Then $\bar{W}(D) = E \rtimes G$ is the split extension of an elementary abelian 2-group E by a group G isomorphic to one of $Sp_{2m}(2)$ $(m \geq 1)$, $O^{\varepsilon}_{2m}(2)$ $(m \geq 3, \varepsilon = \pm)$, or $S_{2m+\delta}$ $(m \geq 2, \delta = 1$ or $2)$. As $\mathbb{F}_2G$-module E is a direct sum of k isomorphic copies of the natural irreducible $\mathbb{F}_2G$-module of dimension 2m.

The assumption that D is connected is not essential. If D is the disjoint union of D_1 (with vertex set X_1) and D_2 (with vertex set X_2) such that no vertex of X_1 is adjacent to any vertex of X_2, then $\bar{W}(D)$ is the direct product of $\bar{W}(D_1)$ and $\bar{W}(D_2)$ acting on $\underline{X}$ as on the direct sum of $\underline{X}_1$ and $\underline{X}_2$. The assumption that D is finite can also be dropped, but the problem then becomes more difficult (see [7]).

The theorem suffers from a failing common to some extent to all the various treatments of the problem. A short list of possibilities for $\bar{W}(D)$ is presented, but the final determination is left somewhat up in the air. We shall discuss this point further and sketch a proof of the theorem in the final section of the paper.

As we have already mentioned, the problem in various forms has surfaced in several different mathematical contexts. (Some of the earlier work dealt only with the important special case of non-degenerate $(\underline{X},f)$.) The problem has appeared in some form in the following areas:

algebraic coding theory:	Ward [22]
algebraic geometry and singularities:	Wajnryb [20], Janssen [14]
finite geometry:	Seidel [18], Hall [6]
group theory:	McLaughlin [17], Fischer [5]
Lie algebras:	Hamelink [8]
surface topology:	Humphries [10], [11], Brown and Humphries [3]

2. MAPPING CLASS GROUPS OF SURFACES

In this section we develop some surface topology in order to show in some detail one occurrence of the problem amid another piece of mathematics. For general reference on the main topics of this section, see [1],[16],[19].

Throughout we denote by T_g a surface of genus g (where, for us, surfaces are 2-manifolds which are connected, orientable, closed and compact). That is, T_g is the surface of a doughnut with g holes. (Throughout this section we assume g>0 unless otherwise stated.)

T_3

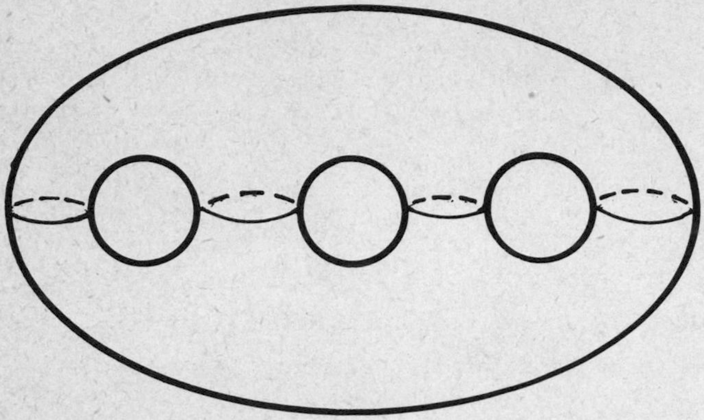

T_3

Let $G^+(T_g)$ be the group of orientation-preserving self-homeomorphisms of T_g. The <u>mapping class group</u> M_g is a homomorphic image of $G^+(T_g)$, factoring out of $G^+(T_g)$ the normal subgroup of all maps homotopic to the identity. (Note that our mapping class group has index 2 in that defined in [19] because we require our maps to be orientation preserving. One frequently takes the equivalent definition of M_g as self-diffeomorphisms of T_g factored by isotopy.) A result of Tietze (see [19]) says that we can also define M_g as the derived group of the outer automorphism group of the fundamental group of T_g:

$$M_g = (\mathrm{Aut}(\pi_1(T_g))/\mathrm{Inn}(\pi_1(T_g)))'.$$

We must take the derived group to preserve orientation. (For any group G, the derived group G' is the normal subgroup generated by all elements $g^{-1}h^{-1}gh$, $g,h \in G$; the factor G/G' is the abelianisation of G.) As the first homology group $H_1(T_g, \mathbb{Z})$ is just the abelianisation of $\pi_1(T_g)$, this last definition of M_g exhibits $H_1(T_g,\mathbb{Z}) \cong \mathbb{Z}^{2g}$ as a $\mathbb{Z} M_g$-module. (Let $a \in \pi_1(T_g)$ map to $\underline{a} \in H_1(T_g,\mathbb{Z})$.)

<u>Example</u> Assume that g = 1 so that we are considering the torus T_1. Let cycles a and b be given as in the diagram (each with a fixed orientation in

order to be considered as an element of $\pi_1(T_1)$).

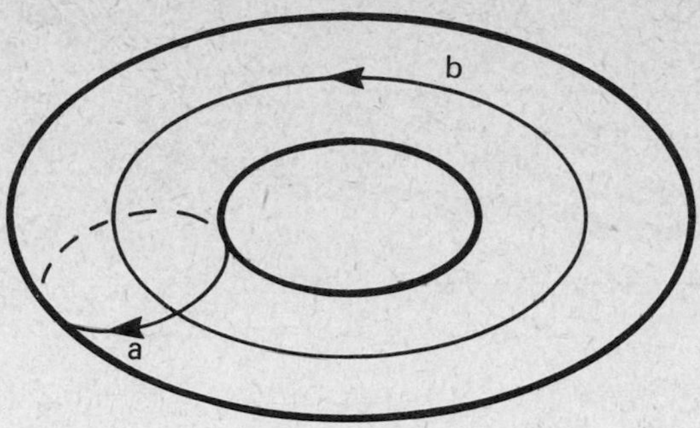

In this case life is made simple by the fact that

$$\pi_1(T_1) \cong H_1(T_1,\mathbb{Z}) \cong \mathbb{Z} \times \mathbb{Z} ,$$

a and b representing generators. Consider the automorphisms $x,y \in \mathrm{Aut}(\pi_1(T_1))$ given by

$$x: \begin{cases} a \longmapsto a \\ b \longmapsto ba \end{cases} \qquad y: \begin{cases} a \longmapsto ab^{-1} \\ b \longmapsto b \end{cases}$$

or, in their homology action,

$$x: \begin{cases} \underline{a} \longmapsto \underline{a} \\ \underline{b} \longmapsto \underline{b} + \underline{a} \end{cases} \qquad y: \begin{cases} \underline{a} \longmapsto \underline{a} - \underline{b} \\ \underline{b} \longmapsto \underline{b} \end{cases} .$$

As matrices,

$$x = \begin{bmatrix} 1 & 0 \\ 1 & 1 \end{bmatrix} \qquad y = \begin{bmatrix} 1 & -1 \\ 0 & 1 \end{bmatrix} ,$$

so that $M_1 = \langle x,y\rangle \cong SL_2(\mathbb{Z}) = \mathrm{Aut}\,(\pi_1(T_1))'$. (Note that $xyx = yxy$, so M_1 is a homomorphic image of the smallest non-trivial Artin braid group **v**[16].)

In the example let us examine the action of x on T_1 more carefully. We cut the torus around a, unfold it, and find:

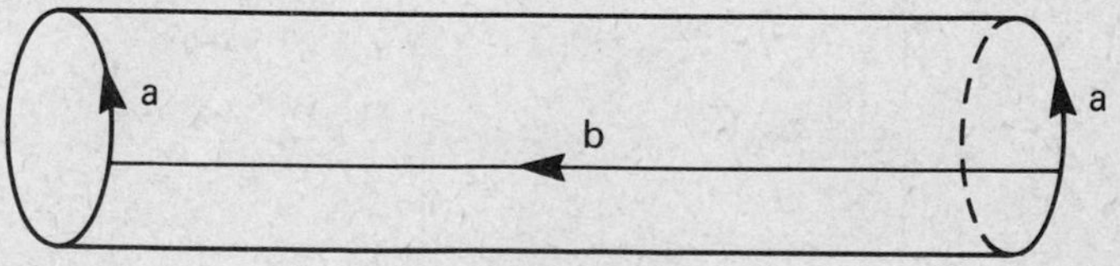

After adding one turn around a to b we have :

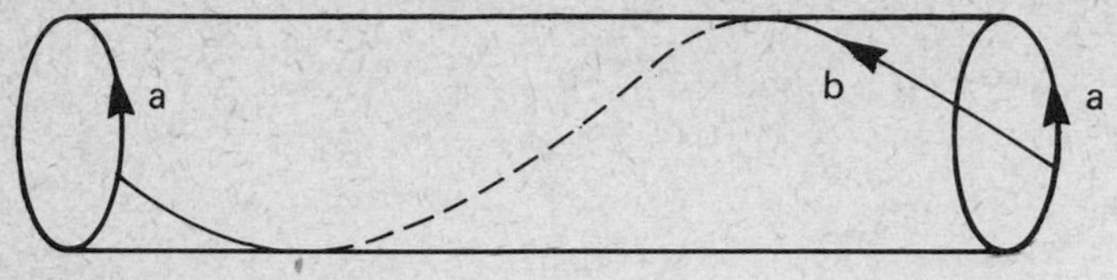

So a homeomorphism representing x can be thought of as being produced by first cutting along a, then adding a complete twist (in the direction determined by a) and finally pasting back together at a.

This construction of a self-homeomorphism is not special to the torus. We can get a homeomorphism of any T_g by cutting along a simple closed curve, twisting once, and pasting together again. If the curve does not bound a disc we produce a non-identity element of M_g, called a <u>Dehn twist</u> after Max Dehn who first studied them in 1938. We call a non-bounding simple closed curve a <u>cycle</u>. It is an easy consequence of the classification of surfaces (see [19]) that there is a unique orbit of cycle classes. In particular, M_g contains a unique conjugacy class of Dehn twists around cycles. Far more difficult is the following result from Dehn's original paper [4].

<u>Theorem</u> For $g>0$ the mapping class group M_g is generated by Dehn twists around cycles. ∎

It is possible in an elementary fashion to give additional structure to M_g and its $\mathbb{F}_2$-module $H_1(T_g,\mathbb{Z})$. Let a and b be (oriented) smooth closed curves on T_g, and suppose z is a point at which they meet transversely.

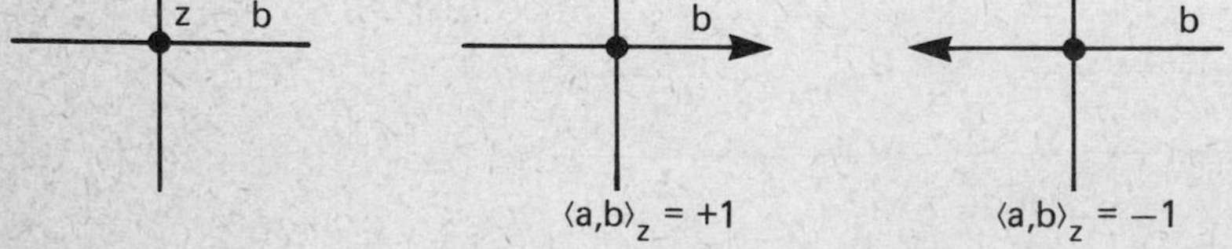

If b moves left-to-right across a, define $\langle a,b\rangle_z = +1$, while if it moves right-to-left across a, define $\langle a,b\rangle_z = -1$. $\langle a,b\rangle$ is then defined as the sum of the $\langle a,b\rangle_z$ over all such z going around a. It is not hard to see that $\langle a,b\rangle$ is well defined and homotopy invariant, and therefore induces a bilinear form (also denoted $\langle .,.\rangle$) on $H_1(T_g,\mathbb{Z})$. This form is called the <u>intersection form</u>. Note that $\langle \underline{a},a\rangle = 0$ for all $\underline{a}\in H_1(T_g,\mathbb{Z})$, because each simple closed curve a is disjoint from some homotopic image of itself (T_g is orientable!). $\langle .,.\rangle$ is therefore a symplectic form on $H_1(T_g,\mathbb{Z})$ which is in fact non-degenerate. ($H_1(T_g,\mathbb{Z})$ has a symplectic basis consisting of cycles.)

<u>Proposition</u> In its representation on $H_1(T_g,\mathbb{Z})$, M_g induces $Sp_{2g}(\mathbb{Z})$. If c is a cycle in T_g, then the Dehn twist around c induces the symplectic transvection

$$\tau(\underline{c}): \underline{v} \longmapsto \underline{v} + \langle \underline{v},\underline{c}\rangle\, \underline{c}. \quad \blacksquare$$

All parts of the proposition are evident from the preceding remarks except the statement that M_g induces all $Sp_{2g}(\mathbb{Z})$ rather than some subgroup; however $Sp_{2g}(\mathbb{Z})$ is generated by its transvections [2, ex. 11, p.87].

As important as M_g is, it is in general difficult to describe. Only $M_0 = 1$ and $M_1 = SL_2(\mathbb{Z})$ have good concrete descriptions. For computational reasons nice presentations of M_g for $g \geqq 2$ are desirable. For M_2 a presentation due to Birman and Hilden in terms of five Dehn twists is given in [1]. The question arises as to whether nice presentations in terms of Dehn twists exist for all g. To find such a presentation it is first necessary to find nice collections of Dehn twists which generate M_g. Dehn [4] gave a set of 2g(g-1) such twists around cycles and Lickorish [15] reduced this number to 3g-1. Humphries [10], [11] then proved that, surprisingly, 2g+1 of the Lickorish twists suffice.

Can M_g be generated by fewer than 2g+1 Dehn twists, and if not, then which collections of 2g+1 twists do generate M_g? It is here that our problem appears on the scene. Let $\mathcal{C} = \{c_1,\ldots,c_n\}$ be a collection of cycles on T_g. To $\mathcal{C}$ we associate the graph $D(\mathcal{C})$, the <u>Dynkin diagram</u> of $\mathcal{C}$, whose vertices are the cycles of $\mathcal{C}$, two such a and b being adjacent precisely when $\langle \underline{a},\underline{b}\rangle$ is odd. We have seen above that a necessary condition for M_g to be generated by the Dehn twists around the cycles of $\mathcal{C}$ is that the transvections induced by $\mathcal{C}$ on $H_1(T_g,\mathbb{Z})$ generate $Sp_{2g}(\mathbb{Z})$.

Theorem (Brown and Humphries [3],[10],[11], Wajnryb [20]) The transvections associated with $\mathcal{C}$ generate $Sp_{2g}(\mathbb{Z})$ if and only if $\bar{W}(D(\mathcal{C}))$ has $Sp_{2g}(2)$ as a homomorphic image. ∎

It is not hard to see that $Sp_{2g}(2)$ is not generated by 2g of its transvections provided g>1:

Corollary (Humphries [10],[11]) For g>1, 2g+1 is the minimal number of Dehn twists generating M_g. ∎

The Humphries/Lickorish generators are the Dehn twists around the cycles shown below:

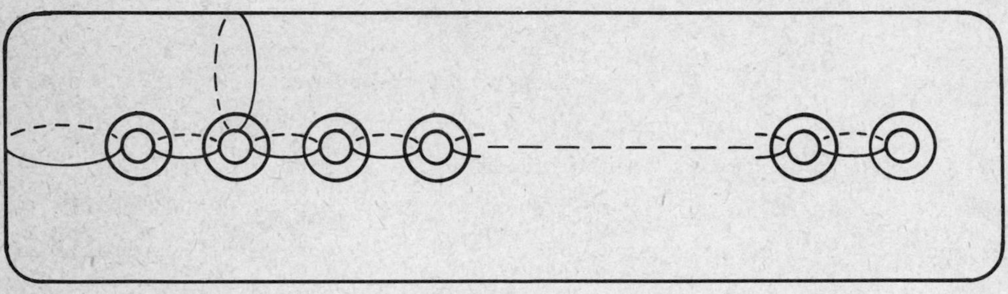

The associated Dynkin diagram is:

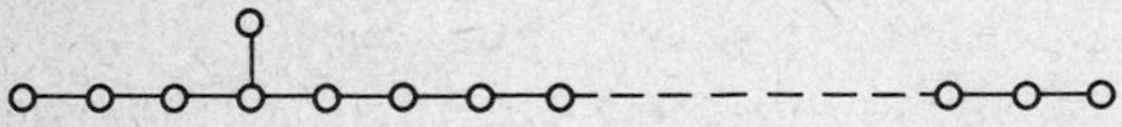

which could be said to have type E_{2g+1}. In fact $\bar{W}(E_{2g+1}) \cong Sp_{2g}(2)$;
for example, $\bar{W}(E_5) \cong W(A_5) \cong S_6 \cong Sp_4(2)$,
$\bar{W}(E_7) \cong Sp_6(2)$.

Using this set of generators, Wajnryb [21] produced presentations for all

M_g, $g \geqq 1$, in a unified and relatively simple form as quotients of the Artin group with associated diagram E_{2g+1}.

We end our excursion into surface topology with a warning. Humphries [11], [12] has proven that the following graph D on $6k + 3 \geqq 15$ vertices satisfies $\overline{W}(D) \cong Sp_{6k+2}(2)$:

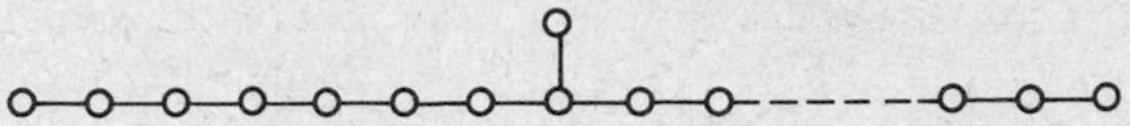

Nevertheless Humphries proved that the set of Dehn twists in M_{3g+1}, analogous to the Humphries/Lickorish generators but with this graph as Dynkin diagram, does not generate M_{3k+1}.

We shall not describe in detail any of the other occurrences of the problem. It is perhaps worth noting that the emergence of $\overline{W}(D)$ as a type of monodromy group in the contexts of complex singularities and projective varieties [14], [20] is similar to the surface topology situation just described. A homology group equipped with an intersection form becomes a symplectic space. (Unlike the mapping class group case this form may be non-degenerate.) Certain distinguished cycles become the centres of the symplectic transvections in the monodromy group. One is then concerned with the types of the symplectic space and of the isometry group generated by the cycles.

3. SOLUTION

We first indicate a proof of the theorem of Section 1 which is similar to the treatment of Ward [22] in that it makes reference to the earlier work of McLaughlin [17] on the non-degenerate case. It is possible to avoid using McLaughlin's results with a small amount of additional effort (see [7]).

Any subspace $\underline{Y}$ of $\underline{X}$ left invariant by the transvection $\tau(\underline{x})$ either contains $\underline{x}$ or is wholly within the subspace $\{\underline{z} \mid f(\underline{x},\underline{z}) = 0\}$. The assumption that D is connected implies easily that all the $\tau(\underline{x})$ with $x \in X$ are conjugate in $\overline{W}(D)$, and therefore any $\overline{W}(D)$-invariant subspace which contains some $\underline{x}$ with $x \in X$ must in fact contain all the $\underline{x}$ with $x \in X$ (note that $g^{-1}\tau(\underline{x})g = \tau(\underline{xg})$). As

$\underline{X}$ is spanned by $\{\underline{x} | x \in X\}$, every proper subspace $\underline{Y}$ of $\underline{X}$ left invariant by $\bar{W}(D)$ must lie within

$$\mathrm{Rad}(\underline{X},f) = \{\underline{z} | (\underline{x},\underline{z}) = 0 \text{ for all } \underline{x} \in \underline{X}\}.$$

Therefore, in its action on the non-degenerate symplectic space $\underline{X}/\mathrm{Rad}(\underline{X},f)$, $\bar{W}(D)$ is irreducible and again generated by symplectic transvections. McLaughlin's theorem applies to say that $\bar{W}(D)$ induces in its representation on $\underline{X}/\mathrm{Rad}(\underline{X},f)$ of dimension 2m the group G isomorphic to one of $Sp_{2m}(2)$, $O^{\varepsilon}_{2m}(2)$, or $S_{2m+\delta}$ (for appropriate $\varepsilon = +$ or $-$ and $\delta = 1$ or 2). We next study the kernel E of this representation. By definition E acts trivially on $\underline{X}/\mathrm{Rad}(\underline{X},f)$, while it acts trivially on $\mathrm{Rad}(\underline{X},f)$ because all of $\bar{W}(D)$ does. Thus, if we extend a basis for $\mathrm{Rad}(\underline{X},f)$ to one for all $\underline{X}$ and write the linear transformations of E as matrices with respect to this basis, we find that E is contained in the subgroup

$$\left\{\left[\begin{array}{c|c} I_{n-2m} & 0 \\ \hline * & I_{2m} \end{array}\right]\right\},$$

a large elementary abelian 2-group. A refinement of this argument reveals what sort of $\mathbb{F}_2G$-module E is. The extension of E by G splits because those matrices in $\bar{W}(D)$ of shape

$$\left\{\left[\begin{array}{c|c} I_{n-2m} & 0 \\ \hline 0 & * \end{array}\right]\right\}$$

constitute a complement to E in $\bar{W}(D)$.

The problem frequently occurs in two other forms which already presume the symplectic setting.

(A) Which vectors of the symplectic space $(\underline{X},f)$ are centres for transvections of $\bar{W}(D)$?

(B) For which symplectic spaces $(\underline{X}',f')$ over $\underline{F}_2$ can one find a spanning subset $\{\underline{x}' | x \in X\}$ such that, for $x,y \in X$,

$f'(\underline{x}',\underline{y}') = 1$ if x and y are adjacent in D
$f'(\underline{x}',\underline{y}') = 0$ if x and y are not adjacent?

Furthermore, what is the isometry group generated by the various $\tau(\underline{x}')$?

As every element of $\underline{X} \setminus \mathrm{Rad}(\underline{X},f)$ is the centre of exactly one symplectic transvection for $(\underline{X},f)$, (A) is effectively equivalent to the original problem (provided both are answered with sufficient precision). A solution to our initial problem yields a solution to (B) because ($\underline{X}$,f) is in a certain sense the "universal solution" for (B). The map

$$T: \underline{x} \longmapsto \underline{x}'$$

extends to a linear transformation

$$T: \underline{X} \longrightarrow \underline{X}'$$

which is an isometry in that

$$f(\underline{u},\underline{v}) = f'(\underline{u}T,\underline{v}T), \quad \text{for all } \underline{u},\underline{v}\in\underline{X}.$$

This forces the kernel of T to lie in Rad($\underline{X}$,f). Conversely, any subspace $\underline{W}$ of Rad ($\underline{X}$,f) provides a symplectic space $\underline{X}/\underline{W}$ which is as in (B).

None of the present solutions is entirely satisfactory. A small list of possible types for $\overline{W}(D)$ is provided by each solution. Sometimes ([6], [22]) no vehicle for deciding the exact type is given, and at best a (conceivably lengthy) process is described which will ultimately lead to a decision ([3], [11],[14]). A better solution would instead decide upon $\overline{W}(D)$ exactly, given a small number of parameters easily calculated from D. (e.g. Humphries [13] has a nice characterisation of those D for which $\overline{W}(D)$ is a symmetric group).Let m and k be the parameters mentioned in the theorem of Section 1, and let $n = |X| = \dim(\underline{X})$. Then we can easily calculate m as

$$2m = \text{row rank } (J).$$

This almost allows the calculation of k, because

$$n - 2m - k = 0 \text{ or } 1.$$

However both cases can occur, and in a specific situation it may be hard to decide which is right. Further, more precise study of the problem is required for a solution of this type. It would also be worthwhile studying in more detail the example in the next-to-last paragraph of the previous section and the other related examples and results due to Humphries [11], [12]. His methods are again largely graphical and may deserve a more general setting.

The problem remains unexhausted. Despite various present solutions, it remains in part unanswered. Conceivably, despite various present applications, its full utility remains undiscovered.

Acknowledgement This paper was written while the author was on sabbatical leave at the Mathematical Institute of Oxford University and was receiving financial support from the NSF and the SERC.

REFERENCES

1. J Birman, Braids, links and mapping class groups, Annals of Math. Studies 82, Princeton University Press 1979.

2. N Bourbaki, Algèbre, Chap. IX, Herman, Paris 1959; MR 21#6384.

3. R Brown and S P Humphries, Orbits under symplectic transvections II, preprint, University College of North Wales 1983.

4. M Dehn, Die Gruppe der Abbildungsklassen, Acta Math. 69(1938), 135-206.

5. B Fischer, Course notes on presentations of 3-transposition groups, 1979-80, unpublished.

6. J I Hall, Linear representation of cotriangular spaces, Lin. Alg. and Appl. 49(1983), 257-273.

7. J I Hall, Graphs, geometry, 3-transpositions, and symplectic F_2-transvection groups, in preparation.

8. R C Hamelink, Lie algebras of characteristic 2, Trans. Amer. Math. Soc. 144(1969), 217-233; MR 40#4321.

9. H Hiller Geometry of Coxeter Groups, Research Notes in Math.54, Pitman, London 1982; MR 83h: 14043.

10. S P Humphries, Generators for the mapping class group, Topology of Low-Dimensional Manifolds, Sussex 1977 (ed. R Fenn), Lecture Notes in Math. 722, Springer-Verlag, New York 1979, 44-47; MR 80i: 57010.

11. S P Humphries, Mapping class groups, graphs, and symplectic geometry, PhD. Thesis, University College of North Wales 1983.

12. S P Humphries, A generalisation of winding number functions on surfaces, preprint, University College of North Wales 1983.

13. S P Humphries, Graphs and Nielsen transformations of symmetric, orthogonal, and symplectic groups, preprint, University College of North Wales, 1984.

14. W A M Janssen, Skew-symmetric vanishing lattices and their monodromy groups, Math. Ann. 266(1983), 115-133.

15. W B R Lickorish, A finite set of generators for the homotopy group of a 2-manifold, Proc. Camb. Phil. Soc. 60(1964), 769-778, and corrigendum 62(1966), 679-681; MR 30#1500.

16. W Magnus, Braid groups: a survey, Proc. Second Int. Conf. Theory of Groups, Canberra 1973 (ed. M F Newman), Lecture Notes in Math. 372(1973), 763-787.

17. J McLaughlin, Some subgroups of $SL_n(F_2)$, Ill.J.Math. 13(1969), 108-115; MR 38#5941.

18. J J Seidel, On two graphs and Shult's characterisation of symplectic and orthogonal geometries over GF(2), T.H.-Report 73-WSK-02, Tech. Univ. Eindhoven 1973.

19. J Stillwell, Classical Topology and Combinatorial Group Theory, Graduate Texts in Math.72, Springer-Verlag, Berlin-New York 1980; MR 82h: 57001.

20. B Wajnryb, On the monodromy group of plane curve singularities, Math.Ann. 246 (1979/80), 141-154; MR 81d: 14016.

21. B Wajnryb, A simple presentation of the mapping class group of an orientable surface, Israel J. Math. 45 (1983), 157-174.

22. H N Ward, Center sets and ternary codes, J. Algebra 65 (1980), 206-224; MR 81m: 94019.

J I Hall,
Department of Mathematics,
Michigan State University,
East Lansing,
Michigan 48824,
USA.

J W P HIRSCHFELD

4 Linear codes and algebraic curves

1. INTRODUCTION

The main object of this paper is to describe a construction by Goppa [5] of a good linear code from a linear series on an algebraic curve over a finite field.

First we give some basic definitions from coding theory. Then we review some results from the classical theory of linear series. This gives enough information for the construction of the codes. This construction can be expressed in terms of more abstract algebraic geometry.

This leads to the question of the maximum number of points on an algebraic curve of genus g. There have been some improvements on the classical result of Hasse and Weil due to Serre [8], which have been partly motivated by the coding theory connection. There is an asymptotic upper bound due to Drinfeld and Vladut [2] which is known to be best possible for square q; this upper bound has a simple and elegant proof.

Finally, we combine an observation in Opencomb [12] with previous results on elliptic curves to obtain arcs which can be completed to arcs larger than those previously known, of size $\frac{1}{2}q + \sqrt{q}$ instead of $\frac{1}{2}q$.

2. LINEAR CODES

MacWilliams and Sloane [10] is followed as a basic reference.

Let $V_{n,q}$ be the vector space $GF(q)^n$ with the usual addition and multiplication. Thus the standard basis is fixed and elements of $V_{n,q}$ are written as row vectors $x = (x_1,\dots,x_n) = x_1 \dots x_n$.

A <u>linear code</u> C is a subspace of $V_{n,q}$ and an element x of C is a <u>codeword</u>. We take dim C = k.

Let $C^{\perp}$ be the subspace orthogonal to C and let H be a matrix whose rows form a basis for $C^{\perp}$. Thus

$$Hx^* = 0$$

for all x in C, where x^* is the transpose of x. Then H is an $r\times n$ matrix of rank $r = n - k$ and is called a <u>parity check matrix</u> of the code C.

The list of symbols is as follows.

n = length of codeword x,

k = dimension of code C,

$r = n - k$ = redundancy of C,

$w(x)$ = weight of x

= number of non-zero coordinates in x,

d = minimum distance of C

$$= \min_{x \in C \setminus \{0\}} w(x) = \min_{x,y, \in C} w(x-y),\ x \neq y,$$

$e = \lfloor \frac{1}{2}(d-1) \rfloor$ = the integral part of $\frac{1}{2}(d-1)$,

$R = k/n$,

$\delta = d/n$.

Since r = rank H, it follows that some r columns of H are linearly independent but no $r+1$ columns are. The definition of d implies that every $d - 1$ columns are linearly independent, but some d columns are linearly dependent. Hence $d - 1 \leqq r$. The code C is called e-error correcting since it can correct e errors.

The usual method of decoding for a linear code is as follows. If y is the received codeword, then y lies in some coset $z + C$ of C, where the coset representative or coset leader z is a vector of minimum weight in the coset. So $y = z + x$ for some x in C and $x = y - z$. Thus y is decoded as x.

Here is an example over GF(2) for which $n = 6$, $k = 2$, $r = 4$, $d = 3$, $e = 1$.

$$H = \begin{bmatrix} 1 & 0 & 0 & 0 & 1 & 0 \\ 0 & 1 & 0 & 0 & 0 & 1 \\ 0 & 0 & 1 & 0 & 1 & 0 \\ 0 & 0 & 0 & 1 & 0 & 1 \end{bmatrix} \qquad C = \begin{bmatrix} 1 & 0 & 1 & 0 & 1 & 0 \\ 0 & 1 & 0 & 1 & 0 & 1 \\ 1 & 1 & 1 & 1 & 1 & 1 \\ 0 & 0 & 0 & 0 & 0 & 0 \end{bmatrix}$$

If $y = 111010$, then y is in the coset $\{111010, 000101, 101111, 010000\}$, whence $z = 010000$ and $x = y - z = 101010$.

The main problem of coding theory is to find codes with large $R = k/n$ for efficiency and large d to correct many errors.

3. ALGEBRAIC CURVES

Let $K = GF(q)$ and $\bar{K}$ be an algebraic closure of K. For x in $V_{n,q} \setminus \{0\}$, define

$$P(x) = \{\lambda x \mid \lambda \in K \setminus \{0\}\}. \tag{3.1}$$

Then $PG(n-1,q) = \{P(x) \mid x \in V_{n,q} \setminus \{0\}\}$ is projective space of n-1 dimensions over GF(q). Subspaces of projective dimension s are denoted Π_s. Let n = 3 and let F in $K[X_0,X_1,X_2]$ be homogeneous of degree M. We define a <u>plane curve</u>

$$\mathcal{C} = V(F) = \{P(x) \mid F(x_0,x_1,x_2) = 0\}. \tag{3.2}$$

The curve $\mathcal{C}$ is <u>absolutely irreducible</u> if F is irreducible over $\bar{K}$, and this is the case we consider throughout. One can think of $\mathcal{C}$ as comprising the affine points P(a,b,1) such that F(a,b,1) = 0 together with the points P(a,b,0) on the ideal line $V(X_2)$ satisfying F(a,b,0) = 0.

The point P(x) is a <u>singular</u> point of $\mathcal{C}$ if

$$F = \frac{\partial F}{\partial X_0} = \frac{\partial F}{\partial X_1} = \frac{\partial F}{\partial X_2} = 0 \tag{3.3}$$

at x in $\bar{K}^3$. Consider the behaviour of P(0,0,1). Let m be the largest value of k such that $F(X_0,X_1,X_2)$ may be expressed as

$$F(X_0,X_1,X_2) = \sum_{i=k}^{M} X_2^{M-i} f_i(X_0,X_1), \tag{3.4}$$

where f_i is homogeneous of degree i. Then P(0,0,1) is a point of <u>multiplicity</u> m on $\mathcal{C}$. So P(0,0,1) is on $\mathcal{C}$ if and only if m > 0 and is singular if and only if m > 1; in the latter case, P(0,0,1) is an <u>ordinary</u> singular point if f_m has no repeated factors. The tangents at P(0,0,1) are given by $V(f_m)$. For example, when

$$F = (X_1^2 - X_0^2)\, X_2 - X_0^3, \tag{3.5}$$

P(0,0,1) is an ordinary double point on $\mathcal{C}$. To see this over the real numbers, plot the points (x,y) such that $y^2 - x^2 - x^3 = 0$.

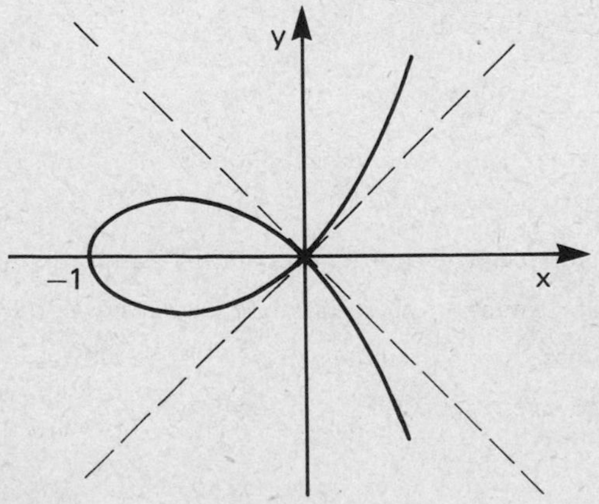

The behaviour of any singular point Q of $\mathcal{C}$ other than P(0,0,1) is obtained by applying a linear transformation taking Q to P(0,0,1). Suppose C has only ordinary singular points $P_1,\ldots,P_t$ of respective multiplicites $m_1,\ldots,m_t$. The <u>genus</u> of C is

$$g = \tfrac{1}{2}(M-1)(M-2) - \tfrac{1}{2} \sum_{i=1}^{t} m_i(m_i-1). \tag{3.6}$$

It has the following properties:

(i) $g \geqq 0$.

(ii) g is invariant under birational isomorphisms; that is, polynomial transformations with a polynomial inverse.

(iii) For a suitable centre Q, a <u>projection</u> π is a birational isomorphism. If V is a curve in PG(r,q) and Π_{r-1} is a hyperplane not containing Q, then $\pi : V \to \Pi_{r-1}$ is given by $P\pi = QP \cap \Pi_{r-1}$ for P in V. Then $V' = \{P\pi\}$ is a curve birationally isomorphic to V.

(iv) When g = 0, C is <u>rational</u>; when g = 1, C is <u>elliptic</u>.

A curve is <u>normal</u> if it is not the projection of a curve of the same order spanning a space of higher dimension. For example, a conic (M = 2) and a non-singular cubic (M = 3, g = 1) in the plane are normal, as is a twisted cubic $T = \{P(s^3, s^2t, st^2, t^3) \mid s, t \in K\}$ in PG(3,q). A singular cubic (M = 3, g = 0) such as (3.5) is not normal, as it is the projection of T.

4. <u>THE CONSTRUCTION OF CODES FROM CURVES</u>

Let K = GF(q), and let C = V(F) be a curve in PG(2,q) consisting of the points $P_1,\ldots,P_n$. Let $G_0,G_1,\ldots G_\rho$ be linearly independent forms in $K[X_0,X_1,X_2]$ of the same degree, such that $F \nmid G$ for each non-zero linear combination

$$G = \lambda_0 G_0 + \ldots + \lambda_\rho G_\rho, \ (\lambda_0,\ldots,\lambda_\rho \in \bar{K}). \tag{4.1}$$

Thus the $(\rho+1) \times n$ matrix $H = [h_{ij}]$ defined by

$$h_{ij} = G_{i-1}(P_j), \ (i = 1,\ldots,\rho+1;\ j = 1,\ldots,n)$$

has linearly independent rows and is therefore a parity check matrix of some code C of codeword length n and with $r = \rho+1$.

A <u>unit</u> code is one in which the sum of the coordinates of each codeword is zero; equivalently, [1 1 ... 1] is a linear combination of the rows of any parity check matrix. Given H as above, we can usually select a linear combination $\phi = \lambda_0 G_0 + \ldots + \lambda_\rho G_\rho$ (where $\lambda_0,\ldots,\lambda_\rho \in K$) such that

$$\phi(P_i) \neq 0, \quad (i = 1,\ldots,n). \tag{4.2}$$

Then the matrix $\tilde{H} = [\tilde{h}_{ij}]$ defined by

$$\tilde{h}_{ij} = G_{i-1}(P_j)/\phi(P_j), (i=1,\ldots,\rho+1;\ j=1,\ldots,n) \tag{4.3}$$

is a parity check matrix of a unit code C, again of codeword length n with $r = \rho+1$.

Example 1

Over $GF(4) = \{0,1,\omega,\omega^2\}$, let $F = X_0^3 + X_1^3 + X_2^3$ and let

$$G_0 = X_0^2,\ G_1 = X_1^2,\ G_2 = X_2^2,\ G_3 = X_0X_1,\ G_4 = X_0X_2,\ G_5 = X_1X_2.$$

Since $x^3 = 0$ or 1 for any x in GF(4), any point on $\mathcal{C} = V(F)$ has one coordinate zero. Hence $n = 9$ and the points are given by the columns of (4.4). The matrix H is given by (4.5).

$$\begin{bmatrix} 0 & 0 & 0 & 1 & 1 & 1 & 1 & 1 & 1 \\ 1 & 1 & 1 & 0 & 0 & \omega^2 & 1 & \omega & 0 \\ 1 & \omega & \omega^2 & 1 & \omega & 0 & 0 & 0 & \omega^2 \end{bmatrix} \tag{4.4}$$

$$H = \begin{bmatrix} 0 & 0 & 0 & 1 & 1 & 1 & 1 & 1 & 1 \\ 1 & 1 & 1 & 0 & 0 & \omega & 1 & \omega^2 & 0 \\ 1 & \omega^2 & \omega & 1 & \omega^2 & 0 & 0 & 0 & \omega^2 \\ 0 & 0 & 0 & 0 & 0 & \omega^2 & 1 & \omega & 0 \\ 0 & 0 & 0 & 1 & \omega & 0 & 0 & 0 & \omega^2 \\ 1 & \omega & \omega^2 & 0 & 0 & 0 & 0 & 0 & 0 \end{bmatrix}. \tag{4.5}$$

Now choose $\phi = X_0X_1 + X_0X_2 + X_1X_2$. Dividing each column of (4.2) by the respective $\phi(P)$, we obtain in (4.6) the matrix $\tilde{H}$.

$$\tilde{H} = \begin{bmatrix} 0 & 0 & 0 & 1 & \omega^2 & \omega & 1 & \omega^2 & \omega \\ 1 & \omega^2 & \omega & 0 & 0 & \omega^2 & 1 & \omega & 0 \\ 1 & \omega & \omega^2 & 1 & \omega & 0 & 0 & 0 & \omega^2 \\ 0 & 0 & 0 & 0 & 0 & 1 & 1 & 1 & 0 \\ 0 & 0 & 0 & 1 & 1 & 0 & 0 & 0 & 1 \\ 1 & 1 & 1 & 0 & 0 & 0 & 0 & 0 & 0 \end{bmatrix}. \tag{4.6}$$

To obtain a basis for the code C is now a standard exercise in linear algebra: we reduce $\tilde{H}$ by row operations to the canonical form $[I_6, A]$, and then the basis of C is given by the rows of

$$B = [A^*, -I_3]$$
$$= [A^*, I_3], \text{ since } K = GF(4).$$

The matrix B is given by (4.7):

$$B = \begin{bmatrix} \omega^2 & 0 & \omega^2 & \omega & \omega & 1 & 1 & 0 & 0 \\ \omega^2 & \omega^2 & 0 & \omega^2 & \omega^2 & 1 & 0 & 1 & 0 \\ \omega^2 & \omega & 1 & \omega^2 & \omega & 0 & 0 & 0 & 1 \end{bmatrix} \tag{4.7}$$

The parameters of C are $n = 9$, $k = 3$, $r = 6$, $d = 6$, $e = 2$.

Suppose instead that ϕ is chosen so that $\phi(P_1) = \phi(P_2) = \ldots = \phi(P_t) = 0$, but $\phi(P_i) \neq 0$ if $i > t$. If the matrix H' obtained by omitting the first t columns of H still has linearly independent rows, then we may proceed as above and again obtain a unit code C, this time with codeword length $n - t$ but still with $r = \rho+1$.

Example 2

Over $GF(q) = \{t_1, \ldots, t_q\}$, let $F = X_2$ and let

$$G_0 = X_0^3, \; G_1 = X_0^2 X_1, \; G_2 = X_0 X_1^2, \; G_3 = X_1^3.$$

The points of $C = V(F)$ are $P(0,1,0)$, $P(1,t_1,0)$, $P(1,t_2,0), \ldots, P(1,t_q,0)$. Thus if we choose $\phi = X_0^3$, then $t = 1$, $\phi(P_i) = 1$ for each remaining point in C, and

$$\tilde{H} = \begin{bmatrix} 1 & 1 & \cdots & 1 \\ t_1 & t_2 & \cdots & t_q \\ t_1^2 & t_2^2 & \cdots & t_q^2 \\ t_1^3 & t_2^3 & \cdots & t_q^3 \end{bmatrix},$$

giving a unit code C with $n = q$, $k = q-4$, $r = 4$, $d = 5$.

In order to investigate how good such codes are in general, we now consider linear series on a curve. This is the subject of the next section.

5. LINEAR SERIES ON A CURVE

Let $C = V(F)$ and $G_0, \ldots, G_\rho$ be as in Section 4. Denote by $\overline{PG}(n,q)$ the projective space of dimension n over $\bar{K} = \overline{GF(q)}$. Let $\bar{C}$ be the curve defined on $\overline{PG}(2,q)$ by F, and let Λ be the following family of curves in $\overline{PG}(2,q)$:

$$\Lambda = \{\bar{G} = V(G) \mid G = \lambda_0 G_0 + \ldots + \lambda_\rho G_\rho,\ \lambda_0,\ldots,\lambda_\rho \in \bar{K}\}.$$

Recall that we assume $F \nmid G$ whenever G is as in (4.1). Then Λ cuts out on $\bar{C}$ a linear series γ_ν^ρ, defined as follows. For each $\bar{G}$ in Λ, the intersection cycle $\bar{C}.\bar{G}$ is the formal sum

$$\bar{C}.\bar{G} = \sum P'_i ,$$

where the P'_i are points that lie on $\bar{C} \cap \bar{G}$. Each distinct point occurs n_i times in the sum, where n_i is the intersection number of $\bar{C}$ and $\bar{G}$ at this point; see Fulton [3, p.74] for a definition of this number. Now, suppose that each of these intersection cycles may be expressed as

$$\bar{C}.\bar{G} = Q_1 + \ldots + Q_t + \sum_{i=1}^{\nu} R_i , \tag{5.1}$$

where $Q_1,\ldots,Q_t$ are fixed and $\sum R_i$ depends on $\bar{G}$ in Λ. Then each such $\sum R_i$ is an element of γ_ν^ρ, a linear series of dimension ρ and order ν. In particular, we may choose $t = 0$, so that $\{\bar{C}.\bar{G} \mid \bar{G} \in \Lambda\}$ is itself a linear series, and the removal of any fixed subset of $\{Q_1,\ldots,Q_t\}$ from each intersection cycle produces another linear series. See Walker [24], for instance, for a more precise formulation. For example, if C is an elliptic cubic and Λ is the pencil of lines through a point P on C, then

$$\{\bar{C}.\bar{G} = P + Q + R \mid \bar{G} \in \Lambda\}$$

is a γ_3^1 on $\bar{C}$. Also

$$\{\bar{C}.\bar{G} - P = Q+R \mid \bar{G} \in \Lambda\}$$

is a γ_2^1 on $\bar{C}$.

A linear series γ_ν^ρ is complete if there is no γ_ν^σ containing γ_ν^ρ. In the above examples, γ_2^1 is complete but γ_3^1 is contained in the γ_3^2 cut out by all lines in the plane.

An adjoint of C is a curve D such that, at any point of C of multiplicity m, the curve D has multiplicity at least $m-1$; further, D is a special adjoint if degree D is $M-3$. The index of speciality i of γ_ν^ρ is the number of linearly independent special adjoints containing an element of γ_ν^ρ. If $i > 0$, then γ_ν^ρ is called special; if $i = 0$, it is non-special.

The series cut out by all special adjoints is the canonical series γ_{2g-2}^{g-1}. For example, if C is a non-singular quartic curve, its genus is 3 and its special adjoints are the lines of the plane, which cut out the canonical γ_4^2.

A series γ_ν^ρ is simple if any element of the series containing an arbitrary point does not necessarily contain another point. Thus a γ_2^1 is not simple.

A curve C is hyperelliptic if it contains a γ_2^1. For example, a quartic with a double point P has a γ_2^1 cut out by the lines through P; its elements are the residual intersections with the curve. For this series, $i = 1$ and $g = \frac{1}{2}(4-1)(4-2) - 1 = 2$.

Lemma 5.1 If $\nu > 2g$, then γ_ν^ρ is simple. ∎

Theorem 5.2 (i) If γ_ν^ρ is complete, then $\nu - \rho \leqq g$.
(ii) (Riemann) If γ_ν^ρ is complete and non-special, then $\nu - \rho = g$.
(iii) (Riemann-Roch) If γ_ν^ρ is complete with index of speciality i, then
$\nu - \rho = g - i$. ∎

Corollary (i) (Clifford) If γ_ν^ρ is special, then $\nu \geqq 2\rho$.
(ii) A complete γ_ν^ρ with $\nu > 2g - 2$ is non-special. ∎

We now construct the projective model of a γ_ν^ρ, as follows. Suppose Λ cuts out γ_ν^ρ on $\bar{C}$ without fixed points. Consider the mapping $\theta\colon \overline{PG}(2,q) \to \overline{PG}(\rho,q)$ defined by

$$\theta(P(x_0,x_1,x_2)) = P(G_0(x_0,x_1,x_2),\ G_1(x_0,x_1,x_2),\dots,G_\rho(x_0,x_1,x_2)).$$

Let $\theta(\bar{C}) = \bar{V}$; then $\theta(\gamma_\nu^\rho)$ is cut out on $\bar{V}$ by hyperplanes and $\bar{V}$ has order ν: that is, a hyperplane of $\overline{PG}(\rho,q)$ meets $\bar{V}$ in ν points.

Theorem 5.3 (i) If γ_ν^ρ on C is complete, then $V = \theta(\bar{C})$ is normal in Π_ρ.
(ii) If γ_ν^ρ is simple, then θ is injective and birational, whence the genus of V is also g.
(iii) If $\nu > 2g$, the curve V is non-singular. ∎

This model of a series gives rise to the following classification of absolutely irreducible curves of genus g.

Theorem 5.4 (i) If $g = 0$, the curve C is birationally isomorphic to the projective line.
(ii) If $g = 1$, the curve C is birationally isomorphic to a non-singular plane cubic.
(iii) A hyperelliptic curve of genus g is birationally isomorphic to a plane curve of degree $g + 2$ with exactly one singular point of multiplicity g.
(iv) A non-hyperelliptic curve of genus $g > 2$ is birationally isomorphic to a projectively unique normal curve of degree $2g - 2$ in Π_{g-1}. ∎

For accounts of the theory of linear series, see Baker [1], Fulton [3], Lefschetz [9], Semple and Kneebone [15], Semple and Roth [16], Severi [19], [20], or Walker [24].

6. THE CONSTRUCTION OF NORMAL CODES

Suppose that the unit code C has been constructed from $C = V(F)$, $G_0,\dots,G_\rho$, and $\phi = \lambda_0 G_0 + \ldots + \lambda_\rho G_\rho$ as in Section 4, and that two further conditions are satisfied:

(a) Λ cuts out a complete γ_ν^ρ on $\bar{C}$; (b) $\nu > 2g$.

Thus $\theta(\gamma_\nu^\rho)$ is cut out on the projective model $\bar{V}$ of $\bar{C}$ by hyperplanes, and $\bar{V}$ is both normal and non-singular. The unit code C is now called a normal code. In both examples 1 and 2, the family Λ satisfies (a) and (b), and therefore produces a normal code.

This construction of codes is not as particular as might seem at first. For, given a parity check matrix H of a code C defined over GF(q), there exists a curve V in PG(ρ,q), $\rho = r-1$, containing the points whose coordinate vectors are the columns of H.

The next proposition uses deep theorems of algebraic geometry to provide results on the parameters of a normal code.

Theorem 6.1 (Goppa) For the normal code C,

(i) $|n - (q + 1)| \leqq 2g\sqrt{q}$;

(ii) $r = \nu - g + 1$;

(iii) $d \geqq \nu - 2g + 2$.

Proof (i) This is just the estimate of Hasse and Weil for the number of points on C.

(ii) By definition $r = \rho + 1$. Since $\nu > 2g$, the series is non-special. Hence, by Theorem 5.2, we have $\rho = \nu - g$, whence the result.

(iii) The minimum weight of C is at least d when every $d-1$ points of V are independent. If we project V from a point P of itself onto a hyperplane not containing P, then this correspondence gives a bijection of V with a normal curve V' of order $\nu - 1$ in $\Pi_{\rho-1}$. We can continue this process while the new curve does not give a special series. So, after t projections, we have a curve $V^{(t)}$ of order $\nu - t$ in $\Pi_{\rho-t}$, which is non-special if $\nu - t \geqq 2g-1$, by Theorem 5.2, Corollary (ii). Thus we may take t as $\nu - 2g + 1$. As $d - 1 \geqq t$, so $d \geqq \nu - 2g + 2$. ∎

How good are these codes?

$$R = k/n = (n-r)/n = (n-\nu+g-1)/n \sim 1-\nu/n + \beta,$$

where $\beta = g/n$. Also

$$\delta = d/n \geqq (\nu - 2g + 2)/n \sim \nu/n - 2\beta.$$

So, for large n,

$$R \geqq 1 - \beta - \delta. \qquad (6.1)$$

The Varshamov-Gilbert bound is given by

$$R \geqq 1 - \Phi(\delta),$$

where

$$\Phi(x) = x \log_q(q - 1) - x \log_q x - (1 - x) \log_q(1 - x).$$

Theorem 6.2 When $\beta < \log_q(2q - 1)-1$, there is an interval (δ_1, δ_2) for which there exist normal codes with higher R than that given by the Varshamov-Gilbert bound.

Proof See [21], [22]. ∎

Projections of codes

The technique of projection both from a point of a curve and from a point off a curve has been used in Sections 3-6. Here we discuss the effect on normal codes.

If V is a curve in Π_ρ, a chordal Π_s is a subspace spanning s + 1 points of V. Let Q be a point not on V and $\Pi_{\rho-1}$ a hyperplane not containing Q. Define the projection $\alpha : V \to \Pi_{\rho-1}$ by $P\alpha = PQ \cap \Pi_{\rho-1}$ and write $V' = V\alpha$. Let the normal code C associated to V have parameters n, d, k, r and the code associated to V' have parameters n', d', k', r'. Then

(i) $r' \leqq r - 1$;

(ii) $n' \leqq n$ with equality if and only if Q is on no bisecant of V;

(iii) $d' \leqq d$ with equality if and only if Q belongs to no chordal Π_{d-2} of V.

This allows the construction of codes with increased n and fixed d.

In Section 6 we have followed the treatment of Goppa [5] very closely. Further details and extensions are given there.

7. THE NUMBER OF POINTS ON AN ALGEBRAIC CURVE

Let C be an absolutely irreducible, non-singular curve of genus g defined over GF(q), and let ν_i be the number of points of C over $GF(q^i)$, $i \geqq 1$. For example, if $C = V(X_0^2 + X_1 X_2)$, then $\nu_i = q^i + 1$. Let

$$N_q(g) = \max \nu_1, \qquad (7.1)$$

where C varies over all curves of fixed genus g. We now describe upper bounds for $N_q(g)$ and in the next sections the exact values known.

The connection between codes and curves described in Section 6 permitted known coding theory bounds to give asymptotic information on $N_q(g)$; see Manin [11]. However, it was then found that classical techniques in number theory gave better results; see Serre [17], [18].

Theorem 7.1 (Hasse-Weil)

$$N_q(g) \leqq q + 1 + \lfloor 2g\sqrt{q} \rfloor. \quad \blacksquare \tag{7.2}$$

Here, as usual, $\lfloor t \rfloor$ denotes the integral part of t.

Corollary For q square and $g = \frac{1}{2}(q - \sqrt{q})$,

$$N_q(g) = q + 1 + 2g\sqrt{q}.$$

Proof The Hermitian curve $\mathcal{U}_2 = V(X_0\bar{X}_0 + X_1\bar{X}_1 + X_2\bar{X}_2)$, where $\bar{X} = X^{\sqrt{q}}$, has $q\sqrt{q} + 1$ points (see [6, p.102]). It is non-singular and of order $\sqrt{q} + 1$, whence its genus $g = \frac{1}{2}(q - \sqrt{q})$ by (3.6). ∎

Theorem 7.2 (Serre [18])

$$N_q(g) \leqq q + 1 + g\lfloor 2\sqrt{q} \rfloor. \quad \blacksquare \tag{7.3}$$

This is an improvement on (7.2). For example, (7.2) and (7.3) respectively give $N_8(2) \leqq 20$ and $N_8(2) \leqq 19$. In fact, $N_8(2) = 18$ (see Section 9).

Theorem 7.3 (Ihara [8])

$$N_q(g) \leqq q + 1 - \tfrac{1}{2}g + \{2(q + \tfrac{1}{8})g^2 + (q^2 - q)g\}^{\frac{1}{2}}. \tag{7.4}$$

Proof Weil showed that

$$\nu_r = q^r + 1 - \sum_{j=1}^{g} (\omega_j^{\,r} + \bar{\omega}_j^{\,r}), \tag{7.5}$$

where $\omega_j \in \mathbb{C}$ and $|\omega_j| = \sqrt{q}$. (For a survey and references, see [7].) Putting $t_j = \omega_j + \bar{\omega}_j$ gives $\omega_j^{\,2} + \bar{\omega}_j^{\,2} = t_j^{\,2} - 2q$. Hence

$$q + 1 - \sum_{j=1}^{g} t_j = \nu_1 \leqq \nu_2 = q^2 + 1 + 2qg - \sum_{j=1}^{g} t_j^{\,2}.$$

Since $g\sum t_j^{\,2} \geqq (\sum t_j)^2$, we have

$$\nu_1 \leqq q^2 + 1 + 2qg - g^{-1}(\nu_1 - q - 1)^2,$$

whence

$$\nu_1^{\,2} - (2q + 2 - g)\nu_1 + (q + 1)^2 - (q^2 + 1)g - 2qg^2 \leqq 0;$$

the result follows. ∎

For $g > \frac{1}{2}(q - \sqrt{q})$, (7.4) gives a better result than (7.2) and (7.3).

To discover what happens for large g, let

$$A_q = \limsup N_q(g)/g. \tag{7.6}$$

Then (7.2) implies that $A_q \leqq 2\sqrt{q}$ and (7.4) implies that $A_q \leqq \frac{1}{2}\{\sqrt{8q+1} - 1\}$.
Note that, in (6.1), the parameter β tends to A_q^{-1} as $n \to \infty$.

<u>Theorem 7.4</u> (Drinfeld and Vladut [2])

$$A_q \leqq \sqrt{q} - 1. \tag{7.7}$$

<u>Proof</u> As in (7.5), with $|\omega_i| = \sqrt{q}$,

$$\nu_1 \leqq \nu_r = 1 + q^r - \sum_{j=1}^{g} (\omega_j^r + \bar{\omega}_j^r). \tag{7.8}$$

To manipulate this formula, define

$$\phi_m(\theta) = (1 + \theta + \ldots + \theta^m)(1 + \theta^{-1} + \ldots \theta^{-m})/(m+1)$$

$$= (1 - \theta^{m+1})(1 - \theta^{-m-1})/\{(1-\theta)(1-\theta^{-1})(m+1)\}.$$

So $\phi_m(e^{i\alpha}) = \{1 - \cos(m+1)\alpha\}/\{(1 - \cos\alpha)(m+1)\}$.

Thus $|\theta| = 1$ implies that $\phi_m(\theta) \geqq 0$.

On the other hand,

$$\phi_m(\theta) = 1 + \sum_{r=1}^{m} c_r (\theta^r + \theta^{-r})$$

where $c_r = (m + 1 - r)/(m + 1)$.

Let $\psi_m(t) = \sum_{r=1}^{m} c_r t^r$;

then

$$\phi_m(t) = 1 + \psi_m(t) + \psi_m(t^{-1}). \tag{7.9}$$

Put $\omega_j = \sqrt{q}\,\theta_j$ so that $|\theta_j| = 1$. Hence

$$\sum_{j=1}^{g} \phi_m(\theta_j) = g + \sum_{j=1}^{g} \{\psi_m(\theta_j) + \psi_m(\theta_j^{-1})\}. \tag{7.10}$$

From (7.8),

$$(\nu_1 - 1)\, q^{-r/2} \leqq q^{r/2} - \sum_{j=1}^{g} (\theta_j^{\,r} + \theta_j^{-r}). \tag{7.11}$$

Multiplying (7.11) by c_r and summing from 1 to m gives

$$\begin{aligned}(\nu_1 - 1)\, \psi_m\, (q^{-\frac{1}{2}}) &\leq \psi_m(q^{\frac{1}{2}}) - \sum_{r=1}^{m} c_r \sum_{j=1}^{g} (\theta_j^{\,r} + \theta_j^{-r}) \\ &= \psi_m(q^{\frac{1}{2}}) + g - \sum_{j=1}^{g} \phi_m(\theta_j) \\ &\leqq \psi_m(q^{\frac{1}{2}}) + g,\end{aligned}$$

where (7.10) and $\phi_m(\theta_j) \geqq 0$ have been used. Therefore

$$\nu_1/g = g^{-1}\, \{1 + \psi_m(q^{\frac{1}{2}})/\psi_m(q^{-\frac{1}{2}})\} + 1/\psi_m(q^{-\frac{1}{2}}). \tag{7.12}$$

However,

$$t \neq 1 \implies \psi_m(t) = \left\{\frac{m}{m+1}\, t - t^2 + \frac{t^{m+2}}{m+1}\right\}/(1 - t)^2\ ;$$

$$|t| < 1 \implies \lim \psi_m(t) = (t - t^2)/(1 - t)^2 = t/(1 - t)$$

$$\implies \lim_{m\to\infty} \psi_m(q^{-\frac{1}{2}}) = 1/(\sqrt{q} - 1).$$

So, given $\varepsilon > 0$, there exists m_0 such that, for $m > m_0$,

$$\psi_m(q^{-\frac{1}{2}})^{-1} < \sqrt{q} - 1 + \tfrac{1}{2}\varepsilon.$$

Then, there exists g_0 such that, for $g > g_0$,

$$g^{-1}\, \{1 + \psi_m(q^{\frac{1}{2}})/\psi_m(q^{-\frac{1}{2}})\} < \tfrac{1}{2}\varepsilon.$$

Thus, for $g > g_0$, (7.12) gives

$$\nu_1/g < \sqrt{q} - 1 + \varepsilon,$$

which is the required result. ∎

Theorem 7.5 ([8], [21], [22]) For q square, $A_q = \sqrt{q} - 1$.

This theorem is proved by considering a curve which parametrizes a class of elliptic curves and showing that $\nu_1/g \geq \sqrt{q} - 1$. ∎

8. ELLIPTIC CURVES

Theorem 8.1 (Waterhouse [25]) For every integer $\nu_1 = q + 1 - t$ with $|t| \leq 2\sqrt{q}$, there exists an elliptic cubic curve in PG(2,q), $q = p^h$, with exactly ν_1 points, provided that one of the following holds:

	Condition on t	Condition on p^h
(i)	$p \nmid t$	
(ii)	$t = 0$	h odd or $p \not\equiv 1 \pmod 4$
(iii)	$t = \pm\sqrt{q}$	h even and $p \not\equiv 1 \pmod 3$
(iv)	$t = \pm 2\sqrt{q}$	h even
(v)	$t = \pm\sqrt{(2q)}$	h odd and p = 2
(vi)	$t = \pm\sqrt{(3q)}$	h odd and p = 3. ∎

Corollary 1 *For* $q = p^h$,

$$N_q(1) = \begin{cases} q + \lfloor 2\sqrt{q} \rfloor, & \text{if } p \text{ divides } \lfloor 2\sqrt{q} \rfloor, h \text{ is odd and } h \geq 3; \\ q + 1 + \lfloor 2\sqrt{q} \rfloor, & \text{otherwise.} \end{cases}$$ ∎

See Table 1 for the values of $N_q(1)$ for small q. The smallest value, and in fact the only value less than 1000, for which the first case of the Corollary applies is $q = 2^7$; see Serre [17]. In [7], it is incorrectly stated that the second case always applies.

Corollary 2 *The number* ν_1 *assumes every integer value in the interval* $q + 1 - \lfloor 2\sqrt{q} \rfloor \leq \nu_1 \leq q + 1 + \lfloor 2\sqrt{q} \rfloor$ if and only if either (i) q = p or (ii) $q = p^2$ with $p \equiv 2$ or $p \equiv 3$ or $p \equiv 11 \pmod{12}$. ∎

A k-arc in PG(2,q) is a set of k points no three of which are collinear. In Opencomb [12], a new class of k-arcs is constructed by P. Neumann. Let $\mathcal{C}$ be an elliptic cubic curve in PG(2,q) with an inflexion O. The points of $\mathcal{C}$ form an abelian group with an identity O as follows; see Walker [24,p.191] for more details. Given P, Q on C, the line PQ meets C again at R; if P = Q, then PQ is the tangent at P. The line OR meets C again at P + Q. Three points P_1, P_2, P_3 are collinear if and only if $P_1 + P_2 + P_3 = 0$. When $\nu_1 = |C|$ is even, then C has a subgroup S of index two.

Theorem 8.2 $K = C \setminus S$ is a $(\nu_1/2)$-arc.

Proof Since $S < C$ and $C = S \cup K$, so $K = \{P_1 + Q \mid Q \in S\}$ for any P_1 in K. Hence $2P_1 \notin K$, whence $2P_1 \in S$. Thus if P_1, P_2 and P_3 are collinear, with P_1 and P_2 in K, we have $P_2 = P_1 + Q$ for some Q in S and $P_3 = -P_1 - (P_1 + Q) = -2P_1 - Q \in S$. So K is a k-arc with $k = \frac{1}{2}\nu_1$. ∎

A complete k-arc is one not contained in a (k + 1)-arc. The maximum size of a k-arc is k = q + 1 for q odd and k = q + 2 for q even. For q odd, a (q + 1)-arc is a conic. For q even, (q + 2)-arcs have not been classified. One example is a conic plus its nucleus; other examples are of the form $\{P(1,t,t^m) \mid t \in K\} \cup \{P(0,1,0), P(0,0,1)\}$. The known complete k-arcs other than those of maximum size all have size at most $\frac{1}{2}q + c$. (See [6, Chapters 8-10], for more details.) It is interesting that the k-arcs constructed above have at most six points on a conic.

Problem For what values of k is the k-arc K of Theorem 8.2 complete?

Let M_q be the largest value of k for which a k-arc can be constructed as in Theorem 8.2. Then, by Theorem 8.1, Corollary 1, since $M_q = \lfloor \frac{1}{2}N_q(1) \rfloor$,

$$M_q \sim \tfrac{1}{2}q + \sqrt{q}.$$

The exact value of M_q for $q \leqq 128$ is given in Table 1.

9. CURVES OF GENUS AT LEAST TWO

Serre [17], [18] has calculated the value of $N_q(2)$ for all q as well as some values of $N_q(3)$ and $N_2(g)$. The techniques for obtaining good upper bounds for $N_2(g)$ are similar to those of Theorem 7.4.

The number $q = p^h$ is special if h is odd and one of the following holds:

(i) p divides $\lfloor 2\sqrt{q} \rfloor$;

(ii) $q = n^2 + 1$, for some n in $\mathbb{Z}$;

(iii) $q = n^2 + n + 1$, for some n in $\mathbb{Z}$;

(iv) $q = n^2 + n + 2$, for some n in $\mathbb{Z}$.

Write $\{x\} = x - \lfloor x \rfloor$, the fractional part of x.

Theorem 9.1 (Serre)

(i) If q is special, then

$$N_q(2) = \begin{cases} q + 2\lfloor 2\sqrt{q} \rfloor, \text{ if } \{2\sqrt{q}\} > \frac{1}{2}(\sqrt{5} - 1); \\ q - 1 + 2\lfloor 2\sqrt{q} \rfloor, \text{ otherwise.} \end{cases}$$

(ii) If q is not special, then

$$N_q(2) = \begin{cases} 2q + 2, \text{ if } q = 4 \text{ or } 9; \\ q + 1 + 2\lfloor 2\sqrt{q} \rfloor, \text{ otherwise.} \end{cases} \blacksquare$$

In Table 1, the value of $N_q(2)$ are given for $q \leqq 128$. The only special q with $\{2\sqrt{q}\} > \frac{1}{2}(\sqrt{5} - 1)$ are q = 2, 8, 128; the other special q are q = 3, 5, 7, 13, 17, 31, 32, 37, 43, 73, 101.

Finally, in Table 2, we give the known improvements to the theorems of Section 7.

Acknowledgement I am most grateful to M Deza and G H Lawden for assistance with the Russian version of Goppa's paper, and to the editors for their suggestions which led to considerable improvements in the exposition of this paper.

TABLE 1

q	$N_q(1)$	$M_q(1)$	$N_q(2)$	$N_q(3)$	q	$N_q(1)$	$M_q(1)$	$N_q(2)$
2	5	2	6	7	49	64	32	78
3	7	3	8	10	53	68	34	82
4	9	4	10	14	59	75	37	90
5	10	5	12	16	61	77	38	92
7	13	6	16	20	64	81	40	97
8	14	7	18	24	67	84	42	100
9	16	8	20	28	71	88	44	104
11	18	9	24	28	73	91	45	106
13	21	10	26	32	79	97	48	114
16	25	12	33		81	100	50	118
17	26	13	32		83	102	51	120
19	28	14	36		89	108	54	126
23	33	16	42		97	117	58	136
25	36	18	46		101	122	61	140
27	38	19	48		103	124	62	144
29	40	20	50		107	128	64	148
31	43	21	52		109	130	65	150
32	44	22	53		113	135	67	156
37	50	25	60		121	144	72	166
41	54	27	66		125	148	74	170
43	57	28	68		127	150	75	172
47	61	30	74		128	150	75	172

TABLE 2

g	$N_2(g)$	g	$N_2(g)$	g	$N_2(g)$
0	3	8	11	16	16–18
1	5	9	12	17	17–18
2	6	10	12–13	18	18–19
3	7	11	13–14	19	20
4	8	12	14–15	20	19–21
5	9	13	14–15	21	21
6	10	14	15–16	39	33
7	10	15	17	50	40

REFERENCES

1. H F Baker, Principles of Geometry, Volume V, Cambridge University Press, Cambridge, 1933 (Ungar, New York, 1960); MR23#A2112.

2. V D Drinfeld and S G Vladut, On the number of points of an algebraic curve (Russian), Functional Anal. Appl. 17 (1983), 68-69.

3. W Fulton, Algebraic Curves, Benjamin, New York, 1969.

4. V D Goppa, Codes on algebraic curves, Soviet Math. Doklady 24 (1981), 170-172; MR82k:94017.

5. V D Goppa, Algebraico-geometric codes, Math. USSR-Isv. 21 (1983), 75-91.

6. J W P Hirschfeld, Projective Geometries over Finite Fields, Oxford University Press, Oxford, 1979; MR81h: 51007.

7. J W P Hirschfeld, The Weil conjectures in finite geometry, in Combinatorial Mathematics X (ed. L R A Casse), Lecture Notes in Math.1036, Springer-Verlag, Berlin 1983, 6-23.

8. Y Ihara, Some remarks on the number of rational points of algebraic curves over finite fields, J. Fac. Sci. Univ. Tokyo Sect. IA Math. 28 (1981), 721-724.

9. S Lefschetz, Algebraic Geometry, Princeton University Press, Princeton, 1953; MR15-150.

10. F J MacWilliams and N J A Sloane, The Theory of Error-Correcting Codes, North-Holland, Amsterdam, 1977; MR57#5408 a,b.

11. J I Manin, What is the maximum number of points on a curve over F_2?, J. Fac. Sci. Univ. Tokyo Sect. IA Math. 28 (1981), 715-720.

12. W E Opencomb, On the intricacy of combinatorial construction problems, Discrete Mathematics, to appear.

13. B Segre, Forme e geometrie hermitiane, con particolare riguardo al caso finito, Ann. Mat. Pura Appl. 70 (1965), 1-201; MR35#4802.

14. B Segre, Introduction to Galois geometries, Atti. Accad. Naz. Lincei Mem. 8 (1967), 133-236; MR39#206.

15. J G Semple and G T Kneebone, Algebraic Curves, Oxford University Press, Oxford, 1959; MR23#A2111.

16. J G Semple and L Roth, Introduction to Algebraic Geometry, Oxford University Press, Oxford, 1949; MR11-535.

17. J P Serre, Nombres de points des courbes algébriques sur F_q. Séminaire de Théorie des Nombres de Bordeaux, exposé no. 22, 1983.

18. J P Serre, Sur le nombre des points rationnels d'une courbe algébrique sur un corps fini, C.R. Acad. Sci. Paris Ser.1, 296 (1983), 397-402.

19. F Severi, Vorlesungen über Algebraische Geometrie (tr. E. Löffler), Teubner, Leipzig, 1921 (Johnson Reprint Corp., New York, 1968); MR39#6880.

20. F Severi, Trattato di Geometria Algebrica, Volume I, Parte I, Zanichelli, Bologna, 1927.

21. M A Tsfasman, Goppa codes that are better than the Varshamov-Gilbert bound, Problems Inform. Transmission 18 (1982), 163-166.

22. M A Tsfasman, S G Vladut and I Zink, Modular curves, Shimura curves and Goppa codes, better than Varshamov-Gilbert bound, Math. Nachr. 109 (1982), 21-28.

23. J F Voloch, Codes and curves, Eureka 43 (1983), 53-61.

24. R J Walker, Algebraic Curves, Princeton University Press, Princeton, 1950 (Dover, New York, 1962); MR11-387.

25. W C Waterhouse, Abelian varieties over finite fields, Ann. Sci. École Norm. Sup. 2 (1969), 521-560; MR42#279.

J W P Hirschfeld,
Mathematics Division,
University of Sussex,
Falmer,
Brighton, BN1 9QH,
East Sussex.

R PENROSE

5 Pentaplexity*

A Class of Non-Periodic Tilings of the Plane

Some readers may be acquainted with an article by Martin Gardner in the January 1977 issue of Scientific American [1]. In this he described a pair of plane shapes that I had found, called "kites and darts", which, when matched according to certain simple rules, could tile the entire plane, but only in a non-periodic way. The tilings have a number of remarkable properties, some of which were described in Gardner's article. I shall give here a brief account explaining how these tiles came about and indicating some of their properties.

The starting point was the observation that a regular pentagon can be subdivided into six smaller ones, leaving only five slim triangular gaps. (See Fig. 1; this is familiar as part of the usual 'net' which folds into a regular dodecahedron.) Imagine, now, that this process is repeated a large number of times, where at each stage the pentagons of the figure are subdivided according to the scheme of Fig. 1.

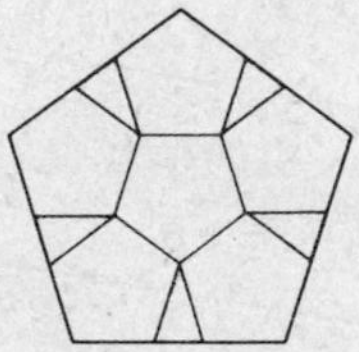

Fig.1

There will then be gaps appearing of varying shapes and we wish to see how best to fill these. At the second stage of subdivision, diamond-shaped gaps appear between the pentagons (Fig. 2). At the third, these diamonds grow 'spikes', but it is possible to find room, within each such 'spiky diamond', for another pentagon, so that the gap separates into a star (pentagram) and a 'paper boat' (or jester's cap?) (Fig. 3). At the next stage, the star and the boat also grow 'spikes', and, likewise, we can find room for new

* We thank the Archimedeans of Cambridge University for permission to reprint Pentaplexity which first appeared in Eureka No.39. The diagrams are reproduced by kind permission of the Mathematical Intelligencer.

pentagons within them, the remaining gaps being new stars and boats (as before). These subdivisions are shown in Fig. 4.

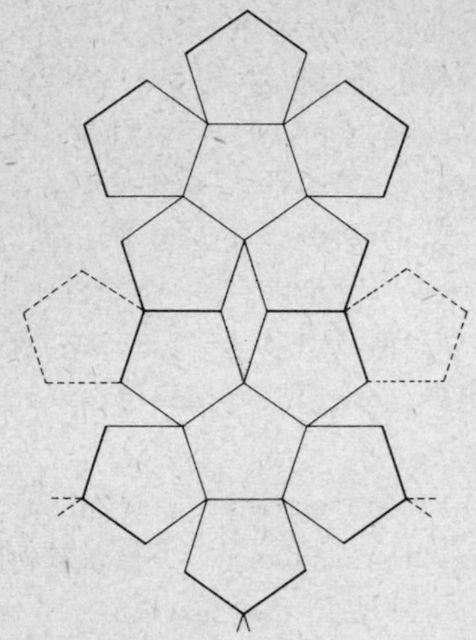

Fig. 2

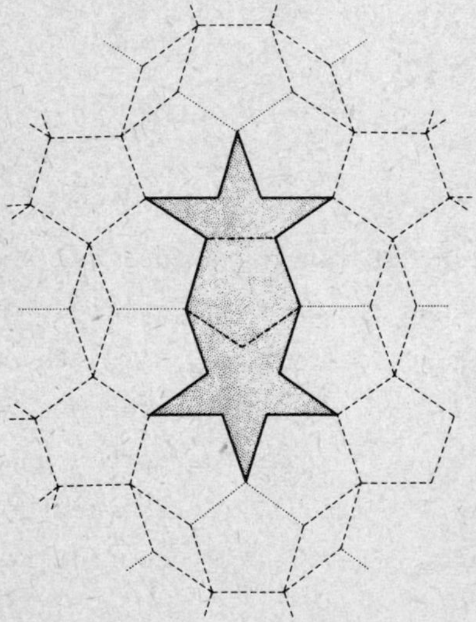

Fig. 3

Fig. 4

Since no new shapes are now introduced at subsequent stages, we can envisage this subdivision process proceeding indefinitely. At each stage, the scale of the shapes can be expanded outwards so that the new pentagons that arise become the same size as those at the previous stage. As things stand, however, this procedure allows an ambiguity that we would like to remove. The subdivision of a 'spiky diamond' can be achieved in two ways, since there are two alternative positions for the pentagon. Let us insist on just <u>one</u> of these, the rule being that given in Fig. 5. (When we examine the pattern of surrounding pentagons we necessarily find that they are arranged in the type of configuration shown in Fig. 5.) It may be mentioned that had the opposite rule been adopted for subdividing a 'spiky diamond', then a contradiction would appear at the next stage of subdivision, but this never happens with the rule of Fig. 5.

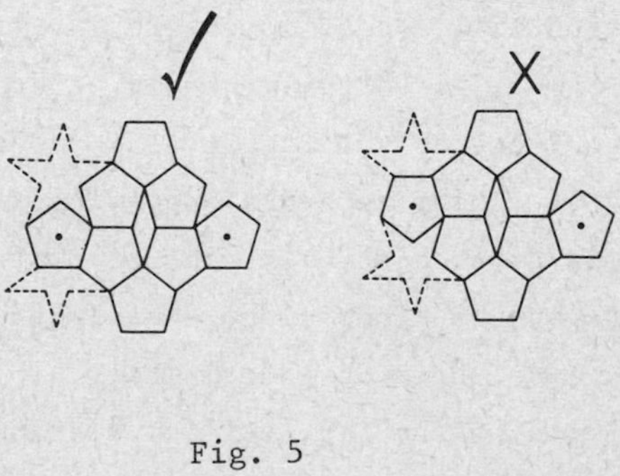

Fig. 5

This procedure, when continued to the limit, leads to a tiling of the entire plane with pentagons, diamonds, boats and stars. But there are many 'incorrect' tilings with these same shapes, being not constructed according to the above prescription. In fact, 'correctness' can be forced by adopting suitable matching rules. The clearest way to depict these rules is to modify the shapes to make a kind of infinite jigsaw puzzle, where a suggested such modification is given in Fig. 6. It is not too hard to show that any tiling with these six shapes is forced to have a hierarchical structure of the type just described.

Fig. 6

Furthermore, the forced hierarchical nature of this pattern implies that the tiling has a number of very remarkable properties. In the first place, it is necessarily non-periodic (i.e. without any period parallelogram). More about this later. Secondly, though the completed pattern is not uniquely determined — for there are $2^{\aleph_0}$ different arrangements — these different arrangements are, in a certain 'finite' sense, all indistinguishable from one another! Thus, no matter how large a finite portion is selected in one such pattern, this finite portion will appear somewhere in every other completed pattern (infinitely many times, in fact). Thirdly, there are many unexpected and aesthetically pleasing features that these patterns exhibit (see Fig. 7). For example, there are many regular decagons appearing, which tend to overlap in places. Each decagon is surrounded by a ring of twelve pentagons, and there are larger rings of various kinds also. Note that every straight line segment of the pattern extends outwards indefinitely, to contain an infinite number of other line segments of the figure. The hierarchical arrangement of

Fig. 7 is brought out in Fig. 8.

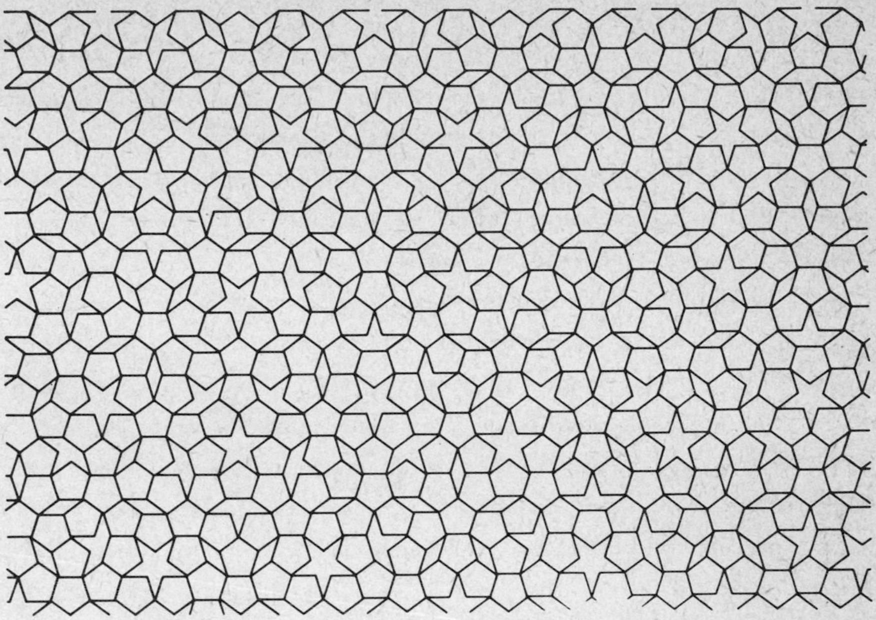

Fig. 7

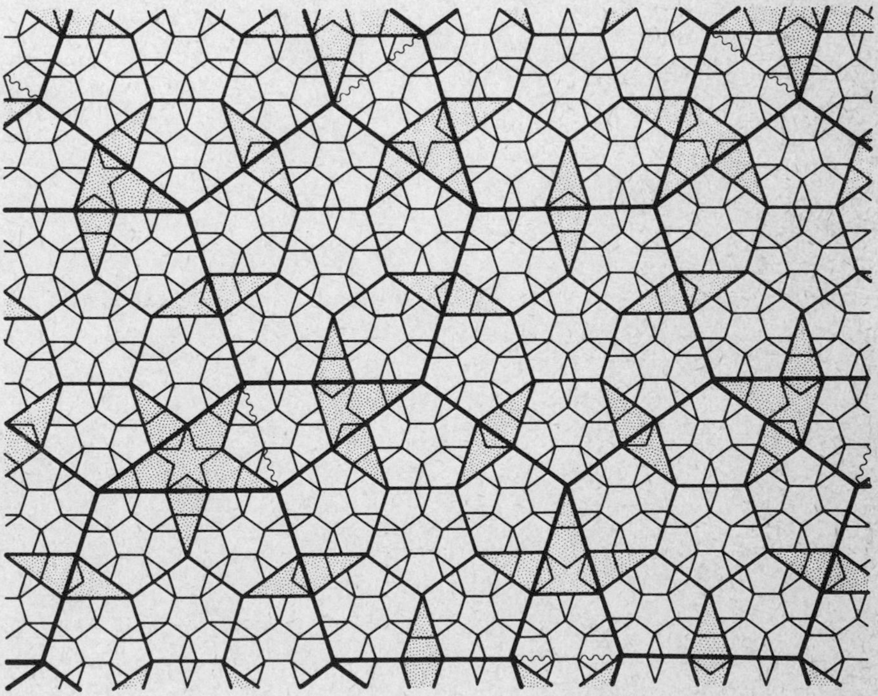

Fig. 8

After I had found this set of six tiles that forces non-periodicity, it was pointed out to me (by Simon Kochen) that Raphael Robinson had, a number of years earlier, also found a (quite different) set of six tiles that forces non-periodicity. But it occurred to me that with my tiles one could do better. If, for example, the third 'pentagon' shape is eliminated by being joined at two places to the 'diamond' and at one place to the bottom of the 'boat', then a set of five tiles is obtained that forces non-periodicity. It was not hard to reduce this number still further to four. And then, with a little slicing and rejoining, to two!

The two tiles so obtained are the 'kites' and 'darts' mentioned at the beginning*. The precise shapes are illustrated in Fig. 9. The matching rules are also shown, where vertices of the same colour must be placed against one another. There are many alternative ways to colour or shade these tiles to force the correct arrangements. One way which brings out the relation to the pentagon-diamond-boat-star tilings is shown in Fig. 10. A patch of assembled tiles (partly shaded in this way) is shown in Fig. 11.

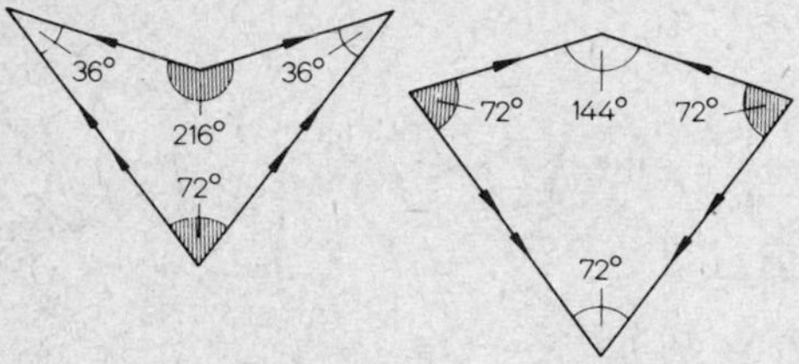

Fig. 9

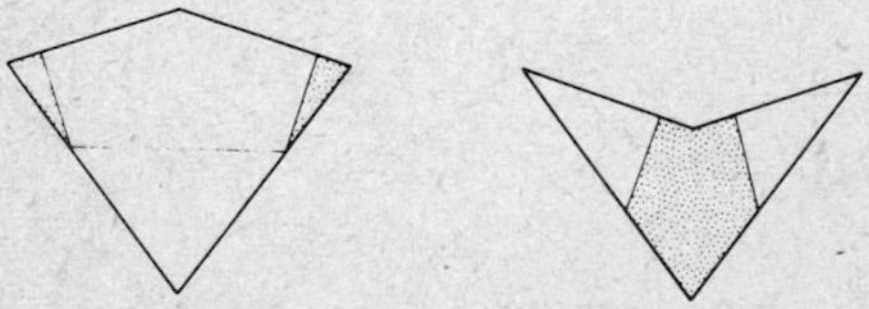

Fig. 10

* These names were suggested by John Conway.

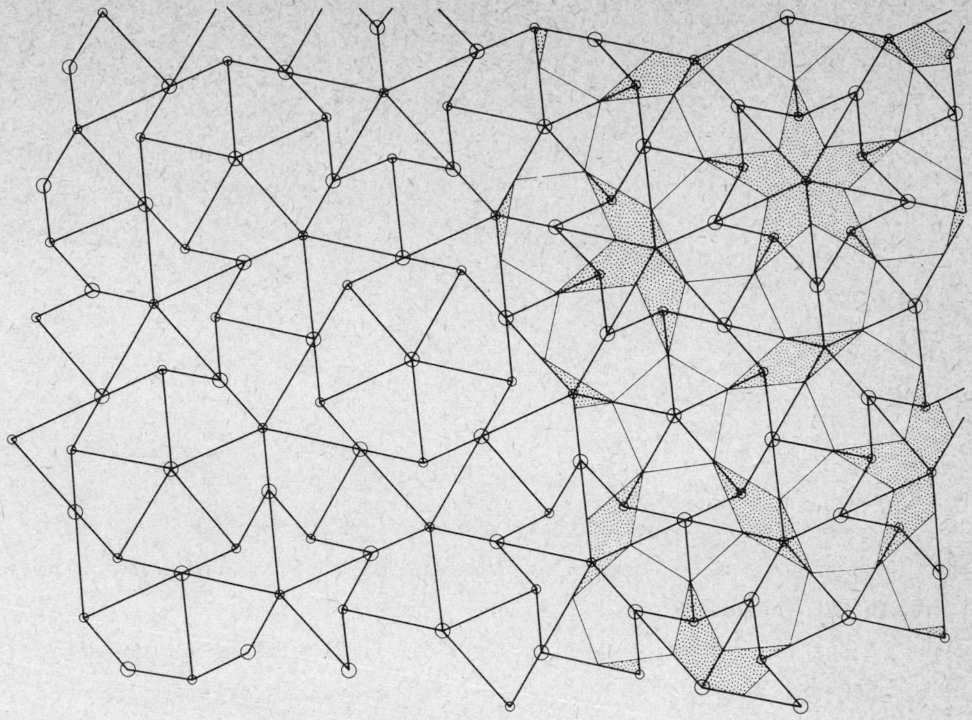

Fig. 11

The hierarchical nature of the kite-dart tilings can be seen directly, and is illustrated in Fig. 12. Take any such tiling and bisect each dart symmetrically with a straight line segment. The resulting half-darts and kites can then be collected together to make darts and kites on a slightly larger scale: two half-darts and one kite make a large dart; two half-darts and two kites make a large kite. It is not hard to convince oneself that every correctly matched kite-dart tiling is assembled in this way. This 'inflation' property also serves to prove non-periodicity. For suppose there were a period parallelogram. The corresponding inflated kites and darts would also have to have the same period parallelogram. Repeat the inflation process many times, until the size of the resulting inflated kites and darts is greater than that of the supposed period parallelogram. This gives a contradiction.

Fig. 12

The contradiction with periodicity shows up in another striking way. Consider a very large area containing d darts and k kites, which is obtained referring to the inflation process a large number of times. The larger the area, the closer the ratio $x = k/d$ of kites to darts will be to satisfying the recurrence relation $x = (1 + 2x)/(1 + x)$ (since, on inflation, a dart and two kites make a larger kite, while a dart and a kite make a larger dart). This gives, in the limit of an infinitely large pattern, $x = \frac{1}{2}(1 + \sqrt{5}) = \tau$, the golden ratio! Thus we get an irrational relative density* of kites to darts — which is impossible for a periodic tiling.

There is another pair of quadrilaterals which, with suitable matching rules, tiles the plane only non-periodically. This is a pair of rhombuses shown in Fig. 13. In Fig. 14 a suitable shading is suggested where similarly shaded edges are to be matched against each other. In Fig. 15, the hierarchical relation to the kites and darts is illustrated. The rhombuses appear mid-way between one kite-dart level and the next inflated kite-dart level.

* This is the numerical density. The kite has τ times the area of the dart, so the total area covered by kites is $\tau^2 (= 1 + \tau)$ times that covered by darts.

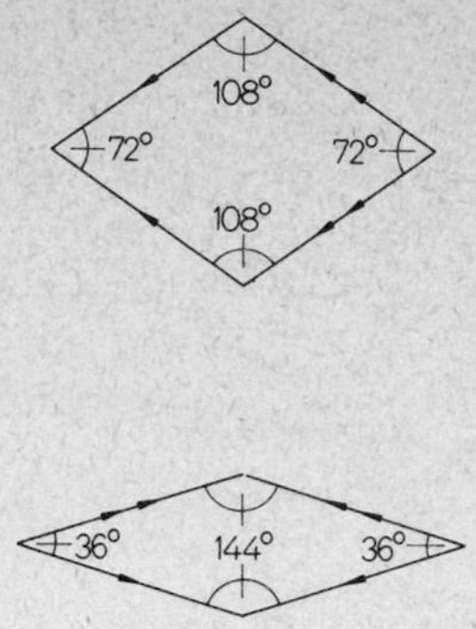

Fig. 13

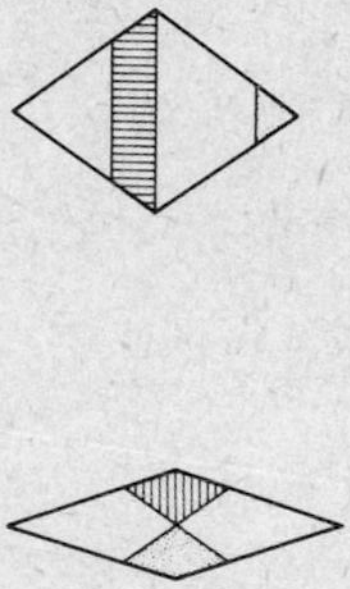

Fig. 14

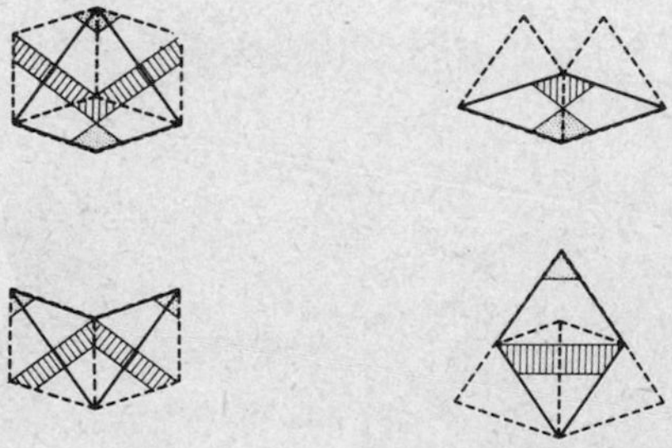

Fig. 15

Many different jigsaw puzzle versions of the kite-dart pair or the rhombus pair can evidently be given. One suggestion for modified kites and darts, in the shape of two birds, is illustrated in Fig. 16. The inflation process (in reverse) is illustrated in Fig. 17.

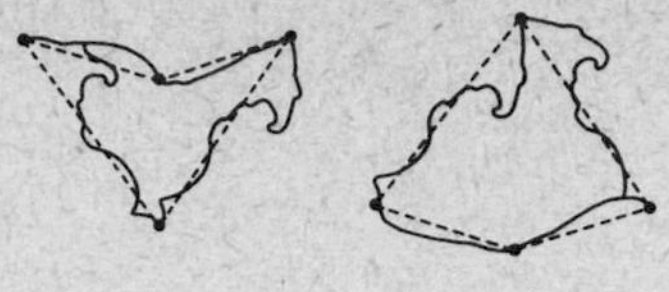

Fig. 16

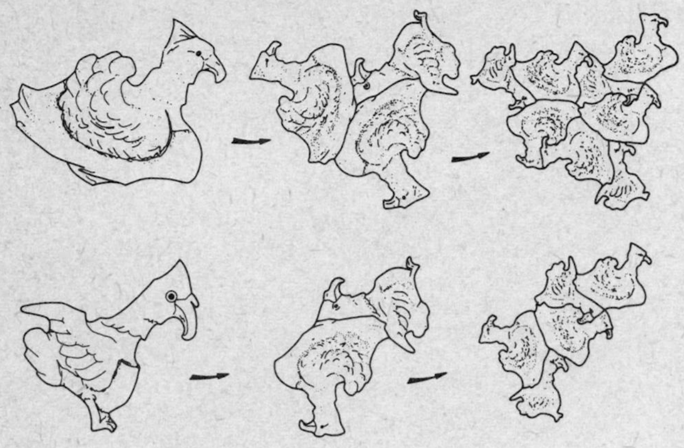

Fig. 17

Other modifications are also possible, such as alternative matching rules, suggested by Robert Ammann (see Fig. 18) which force half the tiles to be inverted.

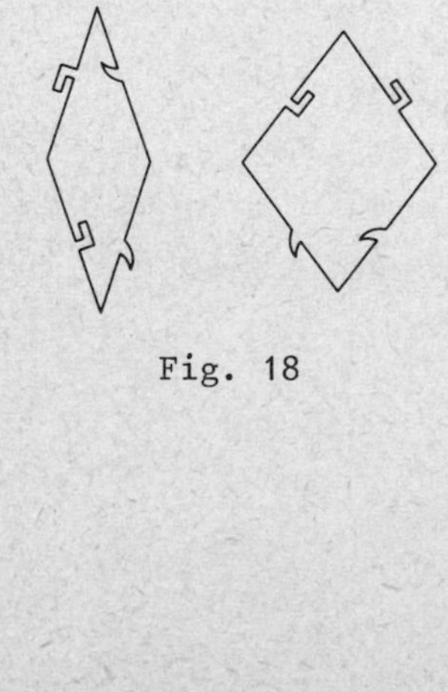

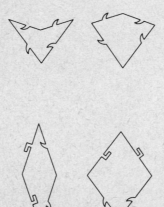

Fig. 18

Many intriguing features of the tilings have not been mentioned here, such as the pentagonally-symmetric rings that the stripes of Fig. 14 produce, Conway's classification of 'holes' in kite-dart patterns (i.e. regions surrounded by 'legal' tilings but which cannot themselves be legally filled), Ammann's three-dimensional version of the rhombuses (four solids that apparently fill space only non-periodically), Ammann's and Conway's analysis of 'empires' (the infinite system of partly disconnected tiles whose positions are forced by a given set of tiles).

It is not known whether there is a single shape that can tile the Euclidean plane only non-periodically. For the hyperbolic (Lobachevski) plane a single shape can be provided which, in a certain sense, tiles only non-periodically (see Fig. 19) — but in another sense a periodicity (in one direction only) can occur. (This remark is partly based on a suggestion of John Moussouris.)

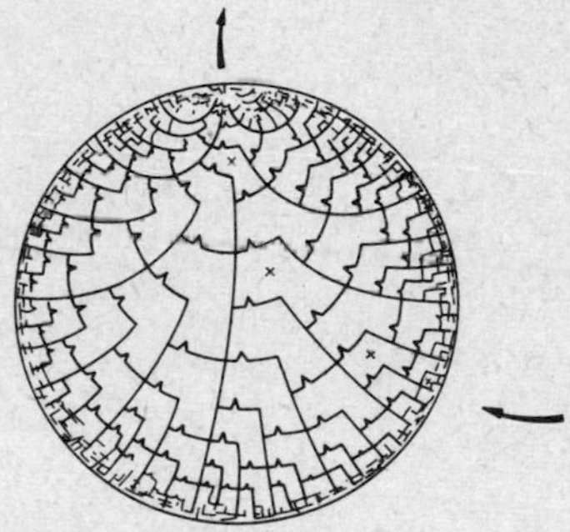

Fig. 19

REFERENCES

1. M Gardner, Mathematical Games, Scientific American, January 1977, 110-121.

2. R Penrose, Bull. Inst. Math. Appl. 10. No.7/8 (1974), 266-271.

R Penrose,
Mathematical Institute,
University of Oxford,
24 - 28 St Giles,
Oxford.

FRED PIPER AND MICHAEL WALKER

6 Binary sequences and Hadamard designs

1. INTRODUCTION

Digital sequences are being used extensively in various communication systems. In particular, random and pseudo-random binary sequences are widely used in applications including synchronization and encryption. In this paper we will see how various desirable properties of the sequences — e.g., long period, balanced statistics and unpredictability — are reflected in the geometric properties of an associated incidence structure. (We will be more specific about these intentions when we have introduced the appropriate terminology.)

Definition A sequence (s_t) is called periodic if there exists an integer m such that $s_{t+m} = s_t$ for $t = 0,1,2, \ldots$.

Clearly if $s_{t+m} = s_t$ for all t then $s_{t+xm} = s_t$ for all t and x. The smallest positive value for m is called the period. An immediate consequence of the definition and the division algorithm is:

Lemma 1 If (s_t) has period p then $s_{t+m} = s_t$ for $t = 0,1,2, \ldots$ if and only if $p|m$.

If (s_t) has period p then, obviously, (s_t) is completely determined by its first p elements. ∎

Definition If (s_t) has period p then any finite subsequence of p consecutive terms is called a cycle. The cycle $s_0, s_1, \ldots, s_{p-1}$ is the generating cycle.

If (s_t) has period p then it has p distinct cycles but, as we have already observed, its generating cycle is unique and completely determines the sequence. Clearly any cycle of length p may be regarded as the generating cycle of a sequence of period p, and the sequences generated by the cycles of (s_t) are closely related to (s_t) itself.

Definition For any sequence (s_t) and any non-negative integer a, the sequence (u_t) defined by $u_t = s_{t+a}$ for t = 0, 1, 2, ... is called a translate of (s_t). The integer a is called the shift.

An immediate consequence of the definition is:

Lemma 2 If (s_t) has period p and if (u_t) is a translate of (s_t) then (u_t) also has period p. Furthermore the cycles of (u_t) are the same as the cycles of (s_t). ∎

If (s_t) is a binary sequence with period p we associate with (s_t) the following $p \times p$ (0,1)-matrix $\underline{A}(s_t)$.

$$\begin{bmatrix} s_0 & s_1 & s_2 & \cdots & s_{p-2} & s_{p-1} \\ s_1 & s_2 & s_3 & \cdots & s_{p-1} & s_0 \\ & \cdot & & & & \cdot \\ & \cdot & & & & \cdot \\ & \cdot & & & & \cdot \\ s_{p-1} & s_0 & s_1 & \cdots & s_{p-3} & s_{p-2} \end{bmatrix}$$

Since $\underline{A}(s_t)$ is a (0,1)-matrix we may regard it as the incidence matrix of an incidence structure $\underline{D}(s_t)$.

Theorem 1 Let (s_t) be a binary sequence with period p. Then :

(i) $\underline{D}(s_t)$ is a square 1-design with a regular cyclic automorphism group;

(ii) if (u_t) is a translate of (s_t) then $\underline{D}(u_t) \cong \underline{D}(s_t)$.

Proof (i) The fact that $\underline{D}(s_t)$ is square with a regular cyclic automorphism group is an immediate consequence of the definition of $\underline{A}(s_t)$. Since the period of (s_t) is p, $\underline{A}(s_t)$ cannot have two identical rows - i.e. $\underline{D}(s_t)$ has no repeated blocks.

(ii) If (u_t) is a translate of (s_t) then, by Lemma 2, the cycles of (u_t) are the same as the cycles of (s_t). Thus, since the rows of $\underline{A}(s_t)$ are the cycles of (s_t), $\underline{A}(u_t)$ can be obtained by permuting the rows of $\underline{A}(s_t)$. Hence $\underline{D}(s_t) \cong \underline{D}(u_t)$. ∎

For any binary sequence (s_t) the complement of (s_t), which we denote by $(\bar{s}_t)$, is the sequence obtained by interchanging the 0s and 1s of (s_t). Clearly $(\bar{s}_t)$ has the same period as (s_t). Another obvious fact is:

Lemma 3 For any periodic sequence (s_t), the incidence structure $\underline{D}(\bar{s}_t)$ is the complement of $\underline{D}(s_t)$. ∎

The aim of this paper is to show how various statistical properties of (s_t) imply specific geometric combinatorial properties of $\underline{D}(s_t)$. Most of the properties considered are of the type required for 'good' pseudo-random binary sequences. The paper is mainly expository and most of the basic results are stated in [2]. A more detailed discussion of binary sequences can be found in [5] or [6].

2. PN-SEQUENCES

Before we can introduce PN-sequences we need a few more definitions.

Definition If (s_t) is a binary sequence then a run of length n is a subsequence of n consecutive identical entries which is neither preceded nor followed by the same element.

So, for example, 0011101 starts with a run of 2 zeros followed by a run of 3 ones. It does not contain a run of 2 ones.

Definition A run of zeros is called a gap and a run of ones is usually called a block. However, in order to avoid confusion with the blocks of our designs, in this paper we will refer to a run of ones as a string.

If (s_t) has period p then, for any a satisfying $0 \leq a < p$, let (s_{t+a}) be the translate of (s_t) with shift a and let A(a) be the number of positions in which the generating cycles of (s_t) and (s_{t+a}) have the same entry.

Definition If (s_t) has period p then the autocorrelation function C(a) is defined by $C(a) = (2A(a)-p)/p$. This is often written as $(A(a)-D(a))/p$, where D(a) is the number of positions in which the generating cycles of (s_t) and (s_{t+a}) disagree.

Clearly $C(0) = 1$. If $a \neq 0$ then the autocorrelation is said to be out-of-phase.

In an attempt to quantify some criteria for assessing a given binary sequence's suitability for use as a pseudo-random sequence, Golomb proposed the following randomness postulates for a binary sequence with (long) period p.

G1 If p is even then a cycle contains $\frac{1}{2}p$ zeros. If p is odd then a cycle contains either $\lfloor \frac{1}{2}p \rfloor$ or $\lfloor \frac{1}{2}p \rfloor + 1$ zeros.

G2 In a cycle of length p, $\frac{1}{2}$ of the runs have length 1, $\frac{1}{4}$ of the runs have length 2 and, in general, for each i for which there are at least 2^{i+1} runs, 2^{-i} of the runs have length i. Moreover, for each of these values of i there are equally many gaps and strings.

G3 The out-of-phase autocorrelation is constant.

Definition A sequence which satisfies G1, G2 and G3 is called a PN-sequence.

Note that if (s_t) is a PN-sequence then $(\bar{s}_t)$ is also a PN-sequence.

There are a number of observations which we should make about Golomb's postulates:

(a) One would expect any (long) sequence obtained by tossing a fair coin to 'almost' satisfy them.

(b) They are probably too precise. As we shall see, the set of sequences satisfying them is probably very limited. Nevertheless any pseudo-random sequence will have characteristics very similar to them.

(c) They are not completely independent. However in certain practical situations one of the properties may be more important than the others. (Note, for instance, that G1 and G2 imply that the probability of a 0 is essentially $\frac{1}{2}$. Thus if a sequence satisfying G1 and G2 is modulo-2 added to any other binary sequence the probability of a 0 in the resultant sequence will also be approximately $\frac{1}{2}$.)

(d) There are a number of (slightly different) formulations of G2. The one given here is used in [2].

3. GEOMETRICAL CONSEQUENCES OF GOLOMB'S POSTULATES

For i = 0, 1, 2, ..., p-1, let $\underline{s}_i$ be the $(i+1)^{th}$ row of $\underline{A}(s_t)$ (regarded as a binary vector) and let x_i be the block of $\underline{D}(s_t)$ determined by $\underline{s}_i$. Thus, for each i, $\underline{s}_i$ is a generating cycle for (s_{t+i}).

Definition If $\underline{x}$ and $\underline{y}$ are binary vectors of the same length then:

(a) the weight of $\underline{x}$, denoted by $w(\underline{x})$, is the number of positions in which $\underline{x}$ has a 1;

(b) the Hamming distance between $\underline{x}$ and $\underline{y}$, denoted by $d(\underline{x},\underline{y})$, is the number of positions in which $\underline{x}$ and $\underline{y}$ have different entries.

Note that $d(\underline{x},\underline{y}) = w(\underline{x}+\underline{y})$.

Golomb's postulate G1 merely says that, for $i = 0,1,2,\ldots,p-1$, $w(\underline{s}_i) = \frac{1}{2}p$ if p is even and either $\lfloor \frac{1}{2}p \rfloor$ or $\lfloor \frac{1}{2}p \rfloor + 1$ if p is odd.

<u>Thus Golomb's randomness postulate G1 merely determines the number of points on each block of $D(s_t)$.</u>

Postulate G3 says that the out-of-phase autocorrelation (that is, C(a) with $a \neq 0$) is a constant. This is equivalent to saying that A(a) and D(a) are constants. From the definition, D(a) is the number of positions in which $\underline{s}_0$ and $\underline{s}_a$ disagree which, clearly, is also the number of positions in which $\underline{s}_i$ and $\underline{s}_{i+a}$ disagree, $i=0,1,\ldots,p-1$. Thus, for any i and a, $D(a) = w(\underline{s}_i + \underline{s}_{i+a})$. Thus if C(a) is constant for all $a \neq 0$ then, for all i,j with $i \neq j$, $w(\underline{s}_i + \underline{s}_j)$ is a constant.

If x_i is the block of $\underline{D}(s_t)$ represented by $\underline{s}_i$ then $w(\underline{s}_i + \underline{s}_j) = |x_i \cup x_j \setminus (x_i \cap x_j)|$. So, if we let $\lambda_{ij} = |x_i \cap x_j|$, $w(\underline{s}_i + \underline{s}_j) = 2(k - \lambda_{ij})$, where k is the number of points on a block. Hence if $w(\underline{s}_i + \underline{s}_j) = \mu$, where μ is constant, then $\lambda_{ij} = (2k - \mu)$ that is, λ_{ij} is independent of i and j. This means that if (s_t) satisfies G3 then $\underline{D}(s_t)$ is the dual of a 2-design which, since it has the same number of points and blocks, implies that it is a 2-design (see [4]).

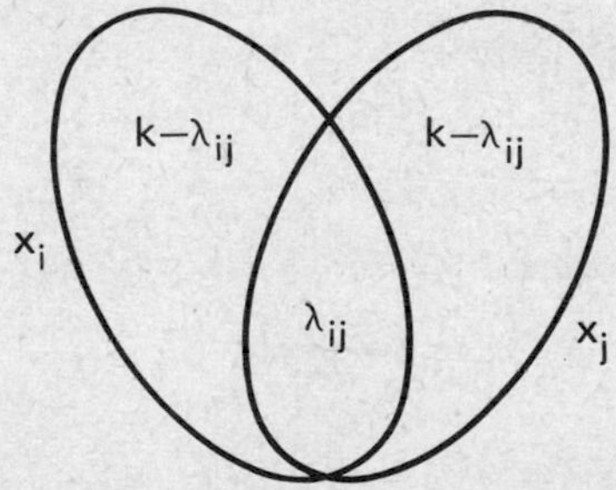

<u>Thus G3 implies that $D(s_t)$ is a symmetric 2-design with a cyclic Singer group.</u>

If $\underline{D}(s_t)$ is a $2\text{-}(p,k,\lambda)$ design then, since it has a cyclic Singer group, the positions of the 1s in the generating cycle of (s_t) must determine a λ-difference set modulo p. As an illustration suppose (s_t) is the sequence of period 7 with generating cycle 0111010. Straightforward verification shows $k = 4$ and $C(a) = -\frac{1}{7}$, $a \neq 0$. It is then easy to check that $\underline{D}(s_t)$ is a symmetric 2-(7,4,2) design and that the set {1,2,3,5}, which is determined by the positions of the 1s in the generating cycle, is a 2-difference set modulo 7.

If (s_t) satisfies G1 and G3 then $\underline{D}(s_t)$ is a symmetric 2-(p,k,λ) design with $p = 2k-1$, $2k$ or $2k+1$. However, for a symmetric 2-design we cannot have $p = 2k$. Thus p must be either $2k-1$ or $2k+1$. From the relation $(p-1)\lambda = k(k-1)$ it is easy to compute λ, and we then have that $\underline{D}(s_t)$ is either a 2-$(4\lambda+3, 2\lambda+1, \lambda)$ design or a 2-$(4\lambda-1, 2\lambda, \lambda)$ design.
So postulates G1 and G3 together imply that $D(s_t)$ is either a cyclic Hadamard 2-design or the complement of a cyclic Hadamard 2-design.

Clearly if $\underline{D}(s_t)$ is a Hadamard design then $\underline{D}(\bar{s}_t)$ is the complement of a Hadamard design. Furthermore, if (s_t) satisfies G1 and G3 then $\underline{D}(s_t)$ is a Hadamard design if a cycle contains $\frac{1}{2}(p+1)$ zeros and $\frac{1}{2}(p-1)$ ones.

It is worth noting here that if $\underline{D}$ is a Hadamard 2-design (or its complement) with a cyclic Singer group, then $\underline{D}$ determines a difference set which, in the way described earlier, gives a periodic sequence satisfying G1 and G3.

The implications of G2 are not as obvious as those of G1 and G3. Indeed the only implication which we shall deduce concerns the total number of runs in a generating cycle. (When counting this number it is absolutely crucial that we consider the cycles as cycles and not as finite sequences of length p. To see why we will consider our earlier example of the sequence generated by 0111010. If we consider this as a sequence of length 7 then it has five runs, whereas the generating cycle of its first translate, that is, 1110100, has only four runs. But if we consider them both as cycles, that is,

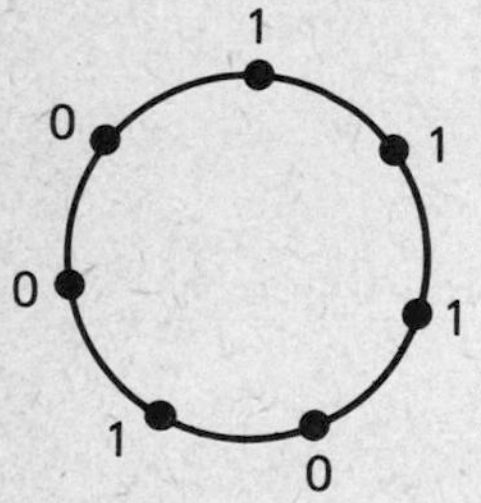

then the number of runs is the same.)

Clearly a generating cycle of (s_t) contains an equal number of gaps and strings. Furthermore this number is equal to the number of occurrences of 01 in $\underline{s}_0$ (where $s_0 = 1$, $s_{p-1} = 0$ is counted). But this is the number of positions in which $\underline{s}_0$ has a 0 and $\underline{s}_1$ has a 1 — that is, it is the number of points which are on the block x_1 but not on x_0. Similarly, the number of gaps in a generating cycle is also equal to the number of occurrences of 10 in $\underline{s}_0$, which is the number of points on x_0 but not on x_1. (Note that since we are only assuming that (s_t) satisfies G2, the incidence structure $\underline{D}(s_t)$ need not be the dual of a 2-design. So this number need not be the same as the number of points on x_0 but not on x_2.) Thus the total number of runs in a generating cycle is $w(\underline{s}_0+\underline{s}_1) = D(1) = \frac{1}{2}p(1-C(1))$.

If we let n be the total number of runs and let i be the largest integer for which there are at least 2^{i+1} runs then, clearly, $2^{i+1} \leq n < 2^{i+2}$. But G2 asserts that 2^{-i} of the runs have length i which, of course, implies $2^i | n$. Thus either $n = 2^{i+1}$ or $n = 3.2^i$.

Postulate G2 also asserts that there are equally many gaps and strings of length i. Since the total number of runs of length i is $2^{-i}n$, this implies that $2^{-i}n$ must be even.

Thus G2 implies that the total number of runs is 2^{i+1}.

The total number of runs with length at most i is

$$(\tfrac{1}{2}+\tfrac{1}{4}+\ldots+2^{-i})2^{i+1} = \sum_{j=1}^{i} 2^j = 2^{i+1}-2.$$

Thus G2 also implies that there are exactly two runs of length greater than i (one gap and one string).

4. THE GEOMETRY OF PN-SEQUENCES

As an immediate consequence of the discussion in the last section we have:

Theorem 2 If (s_t) is a PN-sequence then $\underline{D}(s_t)$ is either a cyclic Hadamard design with parameters $2\text{-}(2^{i+2}-1, 2^{i+1}-1, 2^i-1)$ or the complement of one. ∎

It was widely conjectured that if (s_t) is a PN-sequence then $\underline{D}(s_t)$ is isomorphic to the design of points and hyperplanes of a finite-dimensional projective space over GF(2), or its complement. But recently Cheng [3] has written a thesis on binary sequences. In it he lists a PN-sequence (c_t) with period 127 which (up to complementation) is the only known counter-example to that conjecture. Although the sequence (c_t) was not "known" before Cheng's work, $\underline{D}(c_t)$, which is a Hadamard 2-(127,63,31) design with a cyclic Singer group, was known. In fact in [1] Beaumert lists all such designs. To obtain the sequence (c_t) from [1], we take the difference set with the appropriate parameters labelled (e). From this set we construct the sequence (d_t), where $d_t=1$ in the positions indicated by the difference sets. The sequence (c_t) is then defined by $c_t = d_{39t}$ where 39t is reduced modulo 127.

It is perhaps worth noting here that the converse of Theorem 2 is not true. In other words if $\underline{D}(s_t)$ is a cyclic Hadamard design with parameters $2\text{-}(2^{i+2}-1, 2^{i+1}-1, 2^i-1)$, or the complement of one, then (s_t) need not be a PN-sequence. Thus the existence of a PN-sequence of period 127 is not at all obvious from Baumert's list. In fact Baumert exhibits four non-isomorphic cyclic Hadamard designs on 127 points which do not give PN-sequences.

Cheng has shown, by computer search, that (up to complementation) (c_t) is the only counter-example with $p \leqq 255$.

5. m-SEQUENCES

One popular way of generating periodic binary sequences is by using a shift register with linear feedback.

An *n-stage shift register* consists of n binary storage elements, $S_0, S_1, \ldots, S_{n-1}$, called *stages*, which are connected in series. The contents of the stages change in time with a clock pulse according to the following rule: if $s_i(t)$ denotes the content of S_i after the tth time pulse, then $s_i(t+1) = s_{i+1}(t)$ for $i = 0, 1, \ldots, n-2$ and $s_{n-1}(t+1) = f(s_0(t), s_1(t), \ldots, s_{n-1}(t))$. The function f is called the *feedback function* of the register. If, for every t, we write $s_t = s_0(t)$ then we say that the register *generates* the sequence (s_t). Clearly $s_i = s_i(0)$ for $i = 0, 1, \ldots, n-1$, and the sequence (s_t) is completely determined by $s_0, s_1, \ldots, s_{n-1}$ and the feedback function f.

If $f(s_0(t), s_1(t), \ldots, s_{n-1}(t)) = \sum_{i=0}^{n-1} c_i s_i(t) \pmod 2$, where each c_i is 0 or 1, then the shift register is said to have *linear feedback*. The constants $c_0, c_1, \ldots, c_{n-1}$ are called the *feedback coefficients*. This can be represented by a diagram like the following where $c_i = 1$ stands for a closed connection and $c_i = 0$ stands for an open one.

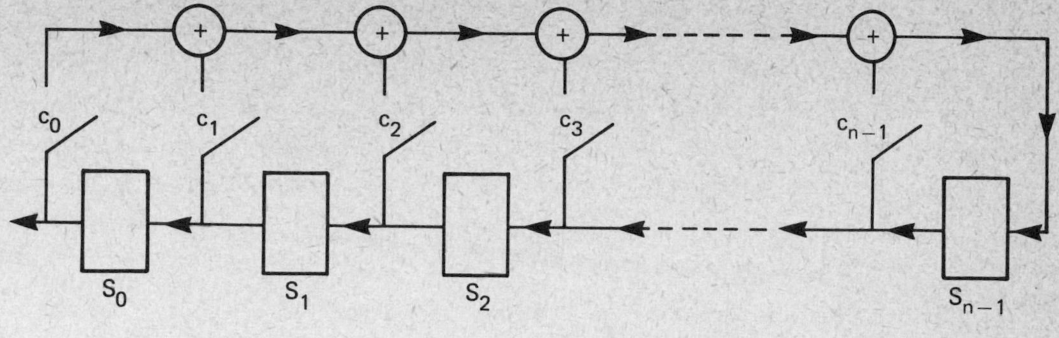

The content of the shift register at any given time is called its state. The following example gives the first 16 state vectors when $n = 4$, $c_0 = c_3 = 1$, $c_1 = c_2 = 0$ and the initial state is 1000.

t	state	t	state
t = 0	1000	t = 8	0101
1	0001	9	1011
2	0011	10	0110
3	0111	11	1100
4	1111	12	1001
5	1110	13	0010
6	1101	14	0100
7	1010	15	1000

Since the state when $t = 15$ is the same as when $t = 0$ the sequence of state vectors must begin to repeat — that is, it has period 15. Similarly the output sequence has period 15 and its generating cycle is 100011110101100.

It is not difficult to see that, in general, the succession of states of an n-stage shift register with linear feedback is periodic with period $p \leqq 2^n-1$. In fact, for any n it is possible to choose the feedback coefficients so that, provided the initial state vector is not zero, the period is equal to 2^n-1. (This is not quite so easy and we refer the interested reader to [2] or [5].)

Definition A sequence of period 2^n-1 which is generated on an n-stage shift register with linear feedback is called an m-sequence.

Historically, m-sequences and PN-sequences have been confused. It was assumed that m-sequences and PN-sequences were the same (up to complementation). But the sequence (c_t) of Cheng provides an example of a PN-sequence which is not an m-sequence.

Any m-sequence is a PN-sequence and, furthermore, if (s_t) is an m-sequence then $\underline{D}(s_t)$ is the complement of a design formed by the points and hyperplanes of projective space over GF(2). (There are many proofs of this. One of the easiest involves showing that the matrix $A(s_t)$ has the correct rank.)

It would be very interesting to know all PN-sequences.

REFERENCES

1. L D Baumert, Cyclic Difference Sets, Lecture Notes in Mathematics 182, Springer-Verlag, Berlin-New York 1971; MR 44#97.

2. H J Beker and F C Piper, Cipher systems: The protection of communications, Van-Nostrand Reinhold, New York 1982.

3. U Cheng, Properties of sequences, Ph.D. thesis, University of Southern California 1981.

4. P Dembowski Finite geometries, Springer-Verlag, Berlin-New York 1968; MR 38#1597.

5. S W Golomb, Shift register sequences (revised edition), Aegean Park Press, California 1982.

6. E S Selmer, Linear recurrence relations over finite fields, Department of Mathematics, University of Bergen 1966.

F C Piper,
Department of Mathematics,
Westfield College (University of London),
Kidderpore Avenue,
London NW3 7ST

M Walker,
Racal Research Ltd.,
Worton Drive,
Worton Grange Industrial Estate,
Reading,
Berks, RG2 OSB.

BRANKO GRÜNBAUM AND G C SHEPHARD

7 The geometry of fabrics

The study of the structure of woven fabrics leads to many interesting problems of a combinatorial or geometric nature. Though there exist many thousands of publications on the practical aspects of weaving and fabric design, there exist less than two dozen papers dealing with the mathematical theory. And this is so in spite of the fact that (as we remarked in [7]) it is difficult to see how mathematicians, and especially combinatorialists, "can fail to be fascinated by the diagrams of fabrics that abound in the literature". In the simplest case (the so-called 2-way 2-fold fabrics) these diagrams consist of rectangular arrays of black and white squares with certain special properties. Clearly such arrays have strong connections with other parts of mathematics; to give just one example, if the black squares are interpreted as 1 and the white squares as -1 we are led to the study of matrices with these two elements, and so to such topics as Hadamard matrices. However fabric designs are, as we shall see, rather special and there are many open problems concerning them, especially in the field of enumeration.

This note is in three parts. The first gives the necessary definitions and terminology. The second is a survey of known results, both published and unpublished. The last part aims to be a complete survey of all the mathematical literature in, and relating to, this area of mathematics.

1. DEFINITIONS AND TERMINOLOGY

Roughly speaking, a fabric consists of several sets of parallel strands which weave under and over each other in some way. More precisely, a strand is a doubly infinite (open) plane strip of constant width. (For purposes of visualisation, think of it as a strip of paper or some other material of negligible thickness.) A layer is a set of parallel strands such that every point of the plane belongs to one of the strands or to the boundaries of two strands. The direction of a layer is that of its strands, so it makes sense to speak of layers being parallel or at some prescribed angle to each other.

By a (2-way 2-fold) fabric we mean two perpendicular layers of strands in the plane E together with a stated ranking at each point of E that does not lie on the boundary of a strand. This means that at every such point (which therefore belongs to precisely one strand of each layer), one layer has precedence over the other. Moreover, this ranking is the same for all points of E that lie in the intersection of two given strands. Because of the obvious practical interpretation we shall often say that the strand of one layer "passes over" the strand of the other layer if it has precedence over it.

There is another condition of importance which we must impose on the ranking of the strands, namely that the fabric must hang together. This means that it is impossible to partition the set of all strands into two non-empty subsets such that every strand of one subset passes over every strand of the other subset. From the point of view of a practical weaver this condition is vital for otherwise the "fabric" will disintegrate into two or more parts. The mathematical investigation of this is the subject of a couple of papers; unfortunately some other authors have disregarded it and consequently claimed the existence of a number of "bogus" fabrics.

We require a representation of a fabric, and for this we use a design (also known as a diagram or draft) that has been traditionally used in the textile industry. This consists of a rectangular array of black or white squares. The rows of squares correspond to weft strands (those that pass from side to side of the fabric as it emerges from the loom) and the columns of squares correspond to warp strands (those that go along the length of the fabric). Each square represents the intersection of a warp strand with a weft strand; it is coloured black if the warp strand passes over the weft strand and white if the weft strand passes over the warp strand.

To illustrate the above definitions we show, in Figure 1, a drawing of (part of) a fabric and also its design. As a mathematical simplification we think of the fabric as extending without bound in all directions — so there are no "edge-effects". If the design of a fabric can be obtained from a suitably coloured $n \times n$ block of squares (a fundamental block) by translations through multiples of n units in horizontal and vertical directions, then we say that the fabric has period n. If, moreover, it is not of period $m < n$ then we say that it has exact period n. We shall be concerned in this note entirely with periodic fabrics.

(a)

(b)

Figure 1. A drawing of a fabric and its design. This fabric has exact period 8.

Figure 1 may be regarded as showing part of a fabric of exact period 8. Any 8 × 8 block of squares selected from the design may be taken as a fundamental block. In Figure 2 we show what purport to be the designs of fabrics — but they are not because in each case the condition that a fabric must hang together is violated. This illustrates the fact that it is often not altogether obvious whether or not this condition is satisfied by a given design.

Figure 2. Designs that do not represent fabrics because the condition that a fabric hangs together is violated.

A <u>symmetry</u> of a fabric F in the plane E is any isometry of E which maps every strand of F into a strand of F and either preserves or reverses all the rankings. The set of symmetries of the fabric F, with the operation of composition, forms the symmetry group of F, denoted by S(F). Symmetries are of two kinds: those that preserve the rankings and are said to preserve the sides of the fabric, and those that reverse all the rankings and are said to interchange the sides of the fabric. The symmetries which preserve the sides of the fabric form a normal subgroup $S_0(F)$ of S(F), and either $S_0(F) = S(F)$ or $S_0(F)$ is of index 2 in S(F). Clearly the symmetry group of a fabric can be read off from the design, but one should carefully note that any symmetry which interchanges warp and weft strands and preserves the sides of the fabric corresponds to a symmetry of the design which interchanges black and white squares.

A fabric F is called isonemal if the symmetry group S(F) is transitive on the strands of F. Curiously, this apparently mathematical concept is of great practical significance; we remark that textbooks on practical fabric design such as Nisbet [16] illustrate many isonemal fabrics. Designs of isonemal fabrics are shown here in Figures 3, 4 and 5. Two familiar classes of fabrics are isonemal. The first of these consists of the twills (Figure 3). The designs are constructed by taking any sequence of black and white squares as one of the rows and then each subsequent row is simply a copy of the previous row shifted one unit (a 1-step) to the right (or to the left). The characteristic feature of a twill is its diagonal "stripe". If the original row of black and white squares repeats with period n, then the whole fabric has period n. Moreover, assuming that the original row contained both black and white squares, then the fabric necessarily hangs together.

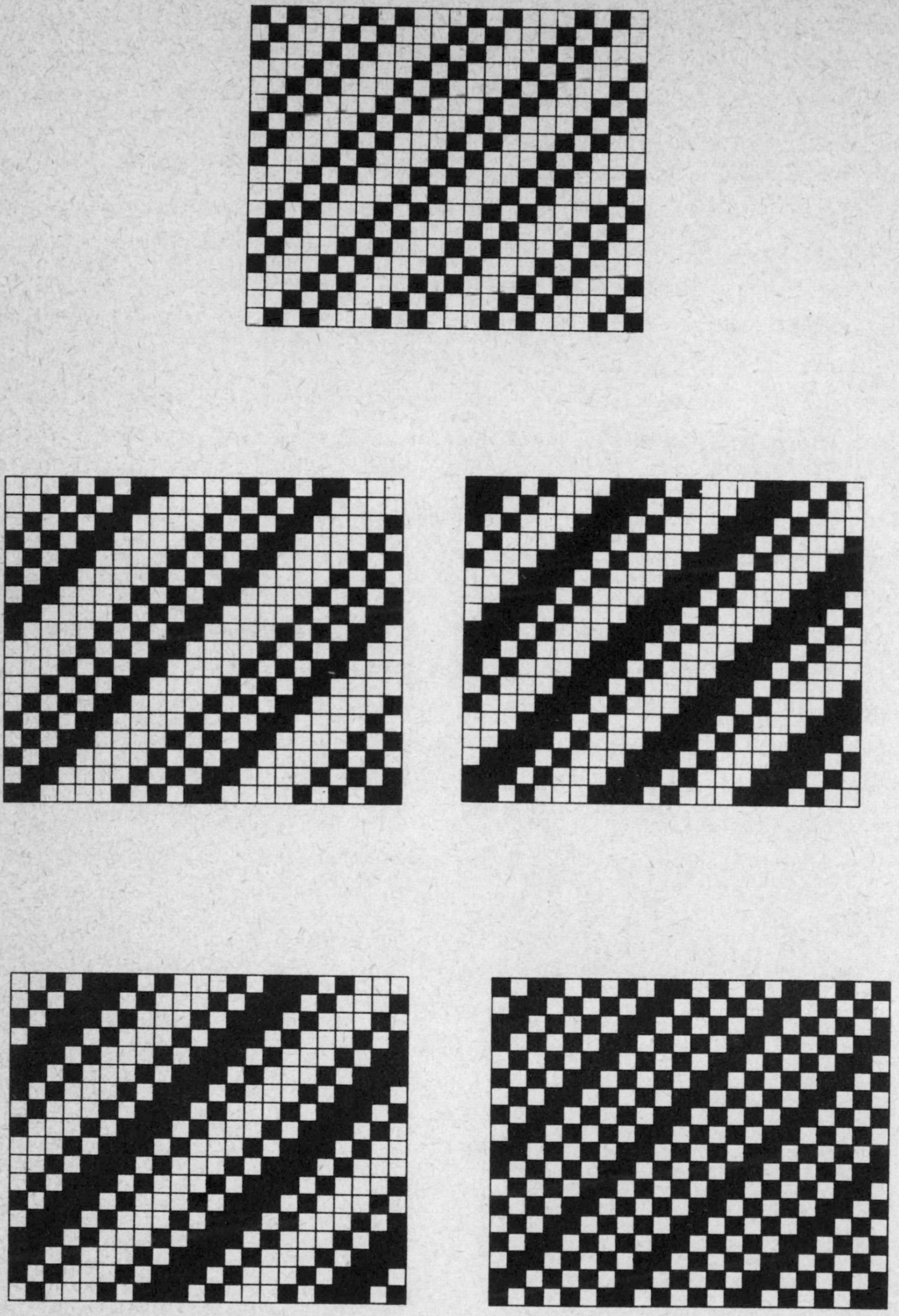

Figure 3. Designs of twills.

The second class consists of the satins. These are isonemal fabrics of exact period n in which each row and each column in a fundamental $n \times n$ block of the design has exactly one black and $(n - 1)$ white squares (or, of course, vice versa), see Figure 4. It will be observed that a fundamental block is constructed by colouring any one square in the top row black, and then subsequent rows are each obtained by an s-step (a translation of s units) to the right from the row above. The satin whose design is obtained in this way is called an (n,s)-satin. Satins are of three kinds: square, rhombic and rectangular. These are shown in Figure 4.

The twills and satins of small period can be enumerated empirically; for larger periods the enumerations become simple exercises in the use of Pólya's Theorem and in number theory. Detailed results are given in [7], and some information is given in Table I on page 88 of this note. In Figure 5 we show some examples of isonemal fabrics which are neither twills nor satins.

An obvious generalisation of the above definitions is to m-way n-fold fabrics. In these, the strands lie in m directions and there are n layers of strands, where $n \geqq m \geqq 2$. (If $n > 2$ then it is possible for several layers of strands to have the same direction.) The ranking of the strands at any point of the plane which does not lie on the boundary of a strand is indicated by a permutation $(i_1, i_2, \ldots i_n)$ which means that the strand from layer i_1 is "topmost", that from layer i_n is "bottommost" and the rest lie in the stated order between these. There is no easy and convenient way of representing such a fabric when $n > 2$; we have to label each region into which the plane is divided by the boundaries of the strands with the corresponding permutation, see Figure 6. However we can supplement this with a sketch showing the direction of the strands in the topmost layer, and this gives some idea of the "appearance" of the fabric. It does not, of course, define it precisely since it gives no information about the order of precedence of the other strands. The symmetry groups of these fabrics are defined in a similar manner to that given above, and, as before, isonemal fabrics are those in which the symmetry group is transitive on the strands.

To a very limited extent 3-way 3-fold fabrics have been used by basket makers and weavers — but no other m-way n-fold fabrics seem to have ever been used in practice.

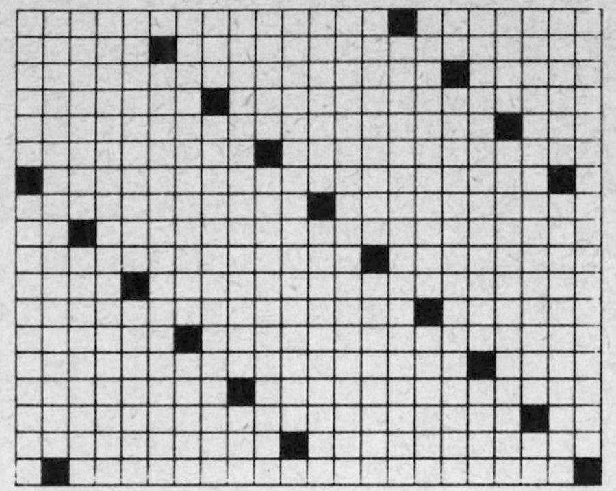

(a) (20, 9)

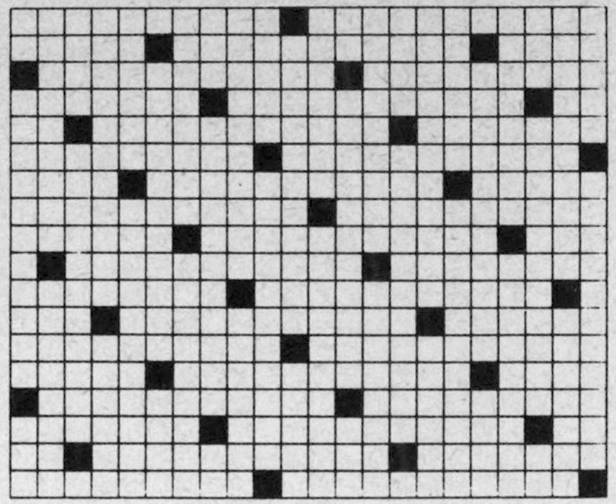

(b) (12, 5)

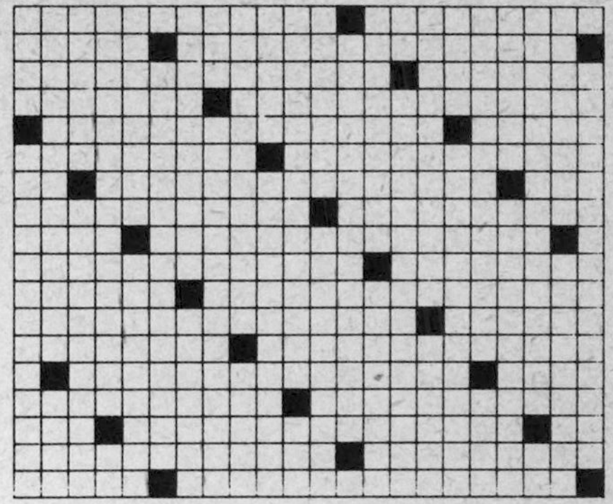

(c) (16, 7)

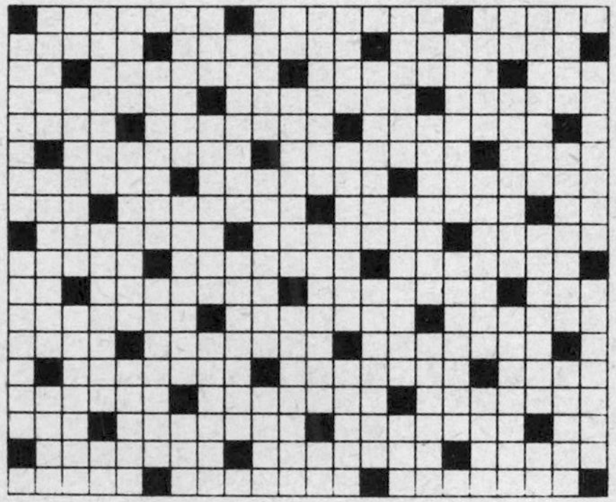

(d) (8, 3)

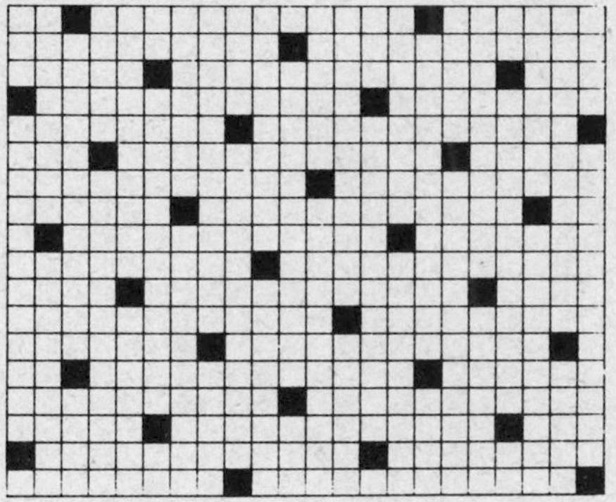

(e) (13, 5)

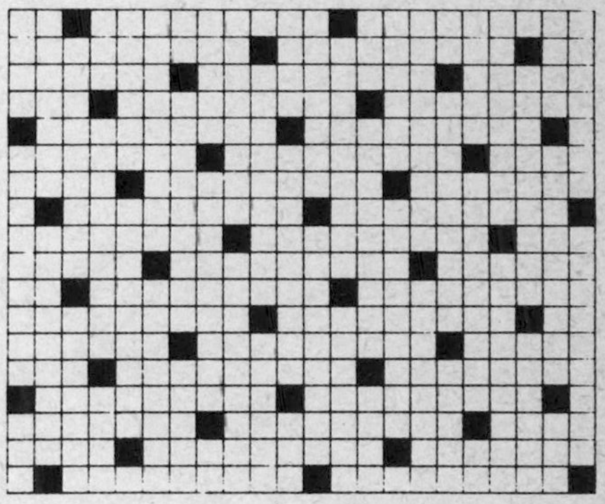

(f) (10, 3)

Figure 4. Designs of satins. (a) and (b) are rectangular satins, (c) and (d) are rhombic satins, (e) and (f) are square satins.

Figure 5. Designs of isonemal fabrics other than satins and twills. The numbers by the designs indicate the genus (or genera) of the fabrics.

1, 3

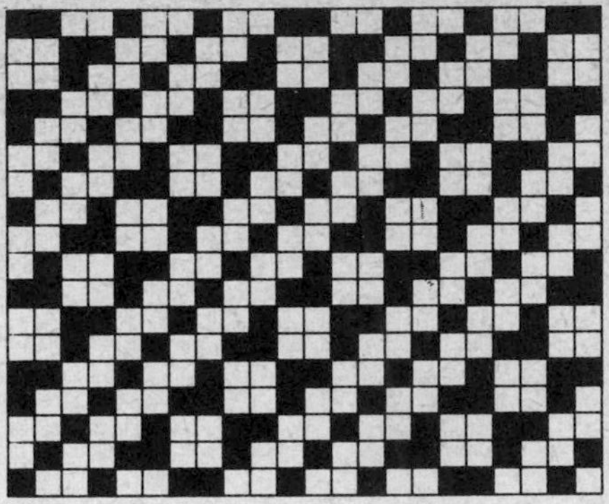

3

3

3

3

3

Figure 5 (continued).

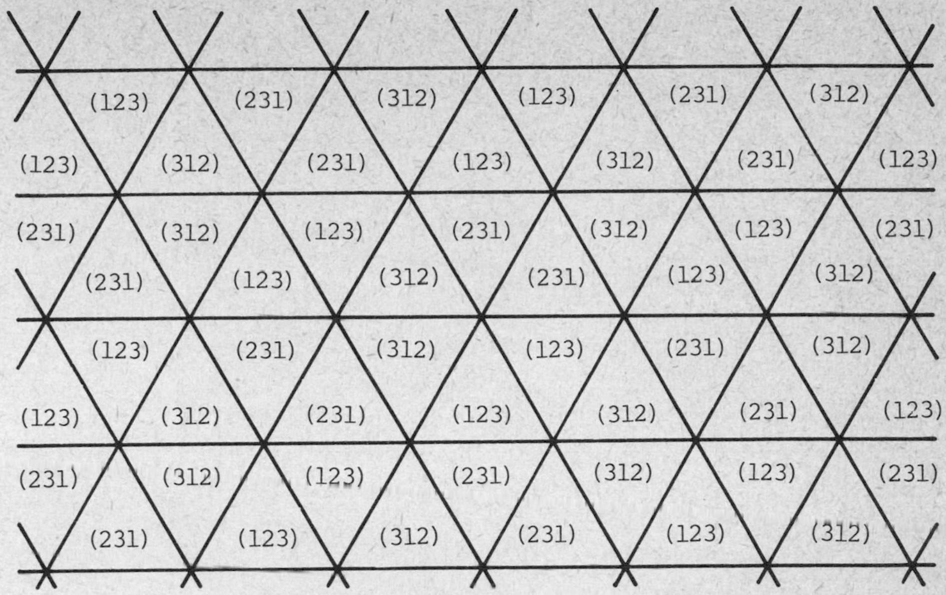

Figure 6 (a). A design for a 3-way 3-fold fabric known as the "mad weave".

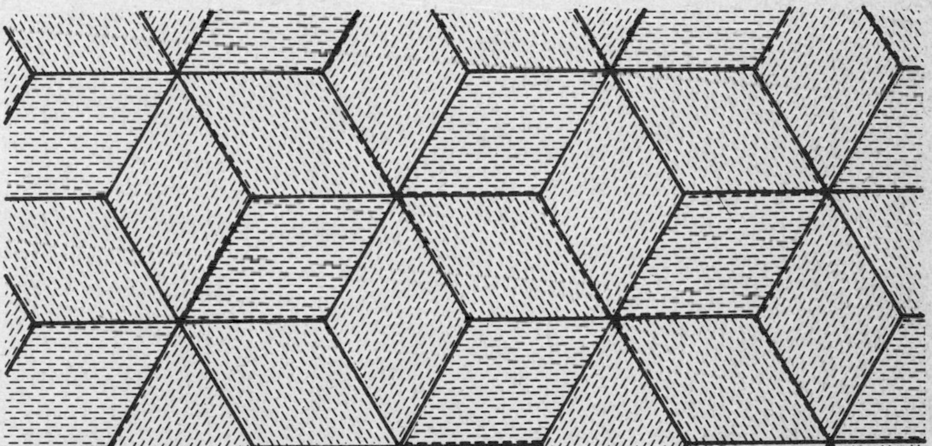

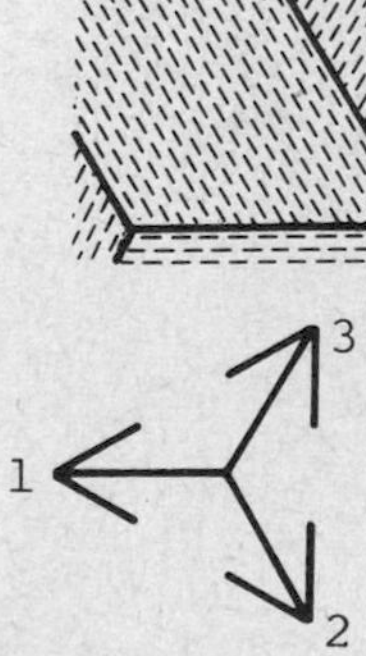

Figure 6 (b) A "sketch" indicating the directions of the strands on the upper side of the "mad weave" fabric of Figure 6 (a).

2. RESULTS AND PROBLEMS

In the case of 2-way 2-fold fabrics the central problem seems to be that of enumeration. As remarked above, satins and twills can be enumerated by standard techniques, but so far the problem of enumerating isonemal fabrics in general have proved intractable. In order to obtain numerical evidence, computer techniques have been used, and it was in this way that the numerical data in Table I was compiled.

TABLE I

The numbers of Isonemal Fabrics

Exact Period	Twills	Satins	Other Isonemal Fabrics	Total Number
2	1	0	0	1
3	1	0	0	1
4	2	0	2	4
5	3	1	0	4
6	5	0	4	9
7	8	0	0	8
8	14	1	40	55
9	21	0	0	21
10	39	1	65	105
11	62	0	0	62
12	112	1	316	429
13	189	1	6	196
14	352	0	*	*
15	607	1	90	698
16	1144	1	*	*
17	2055	1	14	2070
18	3885	0	*	*
19	7154	0	*	*
20	13602	1	*	*

* = unknown at present.

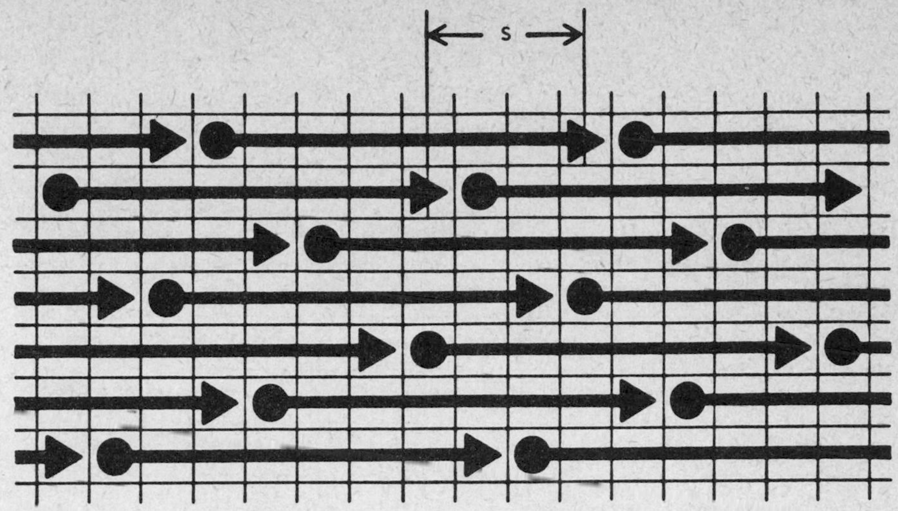

Genus 1.

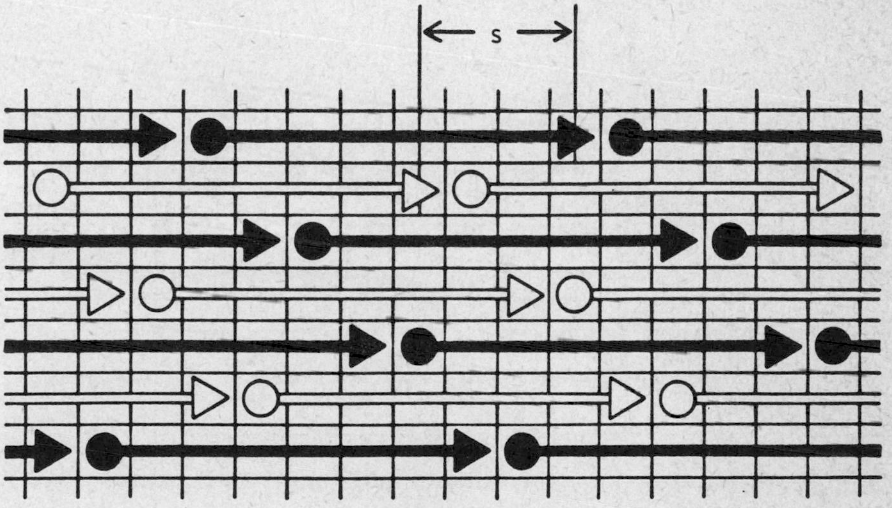

Genus 2.

Figure 7. The five genera of fabrics. A black arrow indicates a sequence of black and white squares; a white arrow indicates the same sequence with the colours black and white interchanged. s and t are parameters.

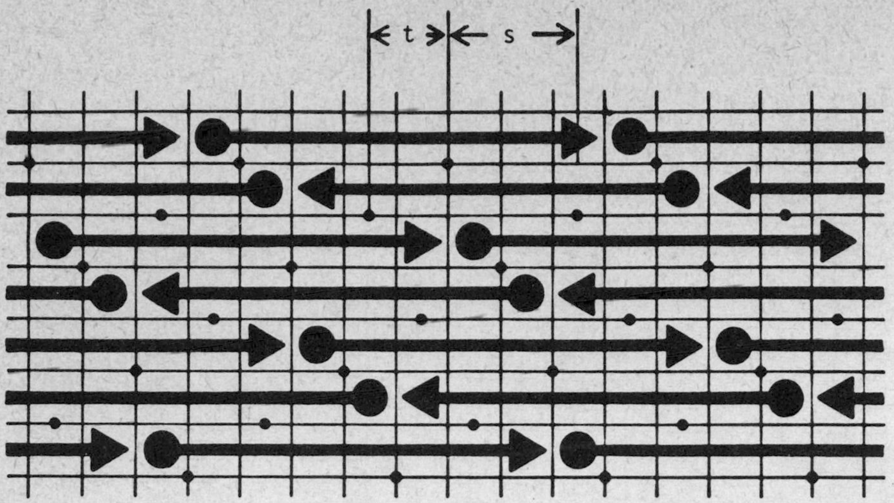

Genus 3.

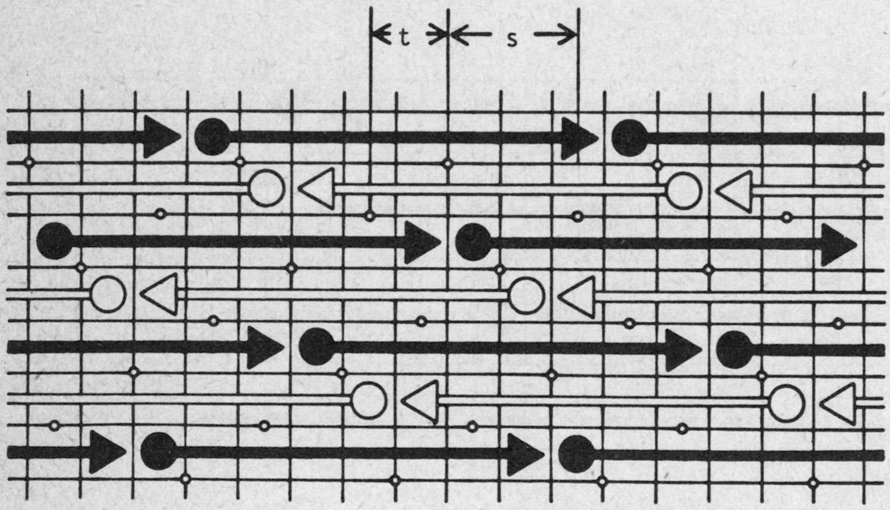

Genus 4.

Figure 7 (continued).

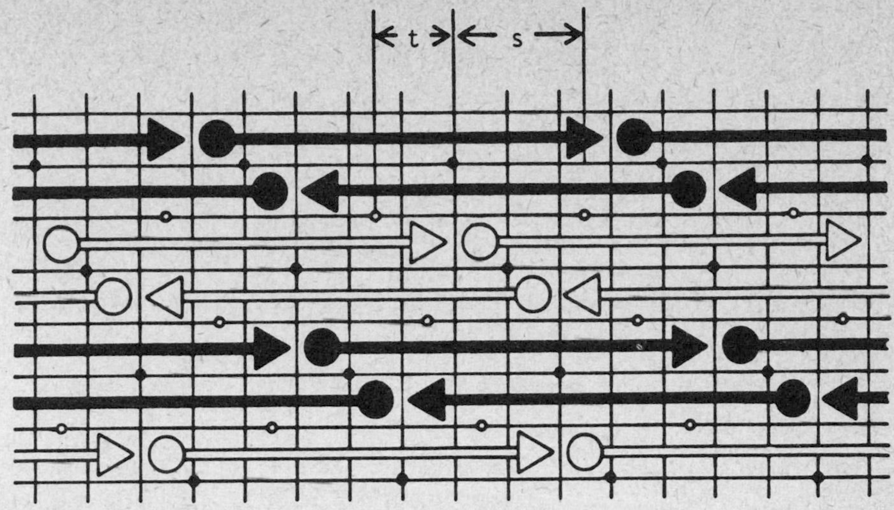

Genus 5.

Figure 7 (concluded).

In view of the difficulty of the general problem, a possible approach seems to be to classify isonemal fabrics in some way and then try to enumerate each of the classes. In the computer enumeration (see [8]) a classification into five genera was used, see Figure 7. Numerical evidence seems to show that the majority of isonemal fabrics of even period are of genus 3 —though we know no explanation of this fact. The "simplest" kind of isonemal fabrics is that of genus 1— here successive rows in the design are produced by applying s-steps to the right, where $1 \leqq s \leqq n - 1$ and s is prime to n. This genus contains, as special cases, the twills and satins, and it seems possible that a theoretical enumeration may be possible in this case. It is worth remarking that designs of genera 1 and 3 always represent fabrics that hang together; those of genera 2, 4 or 5 do not necessarily do so. However, as can be seen from Figure 5, the situation is complicated by the fact that a fabric may belong to more than one genus.

Another way of classifying isonemal fabrics is by properties related to their symmetry groups. In [8] a scheme for doing this on the basis of the appearance of the fabric is proposed. So far as we are aware, a complete enumeration of the possible symmetry groups of fabrics has never been carried out.

In the case of m-way n-fold fabrics the enumeration problems are even more intractable, and even the possible structures of isonemal fabrics are not known. The main theorem in this area is the following.

Theorem If F is an m-way n-fold isonemal fabric, then the pair (m,n) is one of (2,2), (2,4), (4,4), (3,3), (3,6) or (6,6).

The proof of this theorem appears in [10], where fabrics of all these kinds are described. ∎

One way in which to construct 2-way 4-fold, 4-way 4-fold, 3-way 6-fold and 6-way 6-fold fabrics is by a process which we call interlacing. For this, two fabrics (either 2-way 2-fold or 3-way 3-fold) are laid one on top of the other and then some strands in the upper layer of the lower fabric are transposed with strands in the lower layer of the upper fabric at the places where they cross one another. Under certain circumstances (which have never been precisely formulated) this procedure leads to an isonemal fabric. An example of a 2-way 4-fold fabric constructed in this way is shown in Figure 8; similar constructions using 3-way 3-fold fabrics are known but the details are rather more complicated. In fact, infinitely many isonemal fabrics can be produced by interlacing in this way but it is not known whether this procedure yields all 2-way 4-fold, 4-way 4-fold, 3-way 6-fold and 6-way 6-fold fabrics. If other fabrics of these kinds exist, then it would appear to be a comparatively simple task (compared with the other problems that we have stated) to construct such fabrics.

B	A	C	A	B	D	B	A	C	A	B	D	B	A	C	A
A	C	A	B	D	B	A	C	A	B	D	B	A	C	A	B
C	A	B	D	B	A	C	A	B	D	B	A	C	A	B	D
A	B	D	B	A	C	A	B	D	B	A	C	A	B	D	B
B	D	B	A	C	A	B	D	B	A	C	A	B	D	B	A
D	B	A	C	A	B	D	B	A	C	A	B	D	B	A	C
B	A	C	A	B	D	B	A	C	A	B	D	B	A	C	A
A	C	A	B	D	B	A	C	A	B	D	B	A	C	A	B
C	A	B	D	B	A	C	A	B	D	B	A	C	A	B	D
A	B	D	B	A	C	A	B	D	B	A	C	A	B	D	B

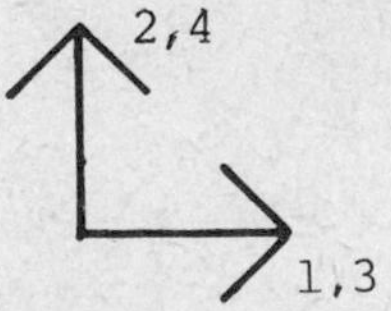

Figure 8. A 2-way 4-fold fabric produced by interlacing two 2-way 2-fold fabrics. This fabric is isonemal.

A: (1234) B: (2143) C: (1324) D: (2413)

It is interesting to note that in the case of 3-way 3-fold fabrics some "partial fabrics", that is parts of fabrics that do not hang together, are used in basketry, see Figure 9. The classification of such fabrics seems to be a totally unexplored area of the subject.

Figure 9. An isonemal 3-way "partial" fabric. This weave is sometimes used by basket makers.

3. LITERATURE AND REFERENCES

The earliest papers on the mathematical theory of fabrics that we have been able to trace are those by Gand [6] and Lucas [19], [20], [21] which appeared at the end of the nineteenth century or in the early part of this century. These were followed, in 1920, by a paper by Shorter [26], and in 1935 by some interesting papers by Woods [28]. In 1978 we published a short abstract [11] about our work on fabrics in which the word "isonemal" was introduced for the first time; this term seems to have now come into general use. More recently a much more general account [7] has appeared, and the reader is referred to this for additional background information. Jean Pedersen [24] extended these ideas to fabrics on polyhedral surfaces; it seems to us that fabrics on bounded or unbounded manifolds (instead of on the plane) present interesting features and would repay further study. For example, on such surfaces the strands can form loops and there exist fabrics with only a finite number (even only two) strands.

In [8] we published a list of all isonemal fabrics with periods up to 13 together with designs of all those of periods up to 8. It is hoped that soon we will be able to supplement this with a paper including designs of all fabrics with periods up to at least 13. All these enumerations were carried out by computer. In [8] we describe a simple procedure for determining whether or not a design represents a fabric that hangs together. This question has also been taken up by Clapham [4] and by Enns [5]. The latter gives a criterion in terms of the connectivity of a bipartite graph.

The only other papers on 2-way 2-fold fabrics are those on the enumeration of special types of fabrics by Hoskins, Street et al. [12], [13], [14], [15], [16], [17], [18]. It is unfortunate that these papers, while asserting that they enumerate fabrics, for the most part only enumerate fabric designs without regard to the important condition that the "fabric" represented must hang together. Some apparently attractive designs (see Figure 2) are therefore useless for the practical weaver. The results in these papers are therefore related to those which are simply concerned with patterns and other arrangements of black and white squares in a rectangular grid, such as Steggall [27], Astle [1], Bouwkamp et al. [3] and the Norwegian squares of Selmer [25].

In the case of m-way n-fold fabrics with $m > 2$ or $n > 2$ the literature is extremely limited. The main theorem stated in the previous section is proved in [10], though it was quoted as early as 1978 in [11] and 1981 in [24]. So far as we know, this is the only mathematical reference in this area, though 3-way fabrics occur in the literature of practical weaving (Mason [22]). Publicity literature from the Barber-Coleman Company of Illinois, USA [2] describes a loom capable of weaving 3-way 3-fold fabrics (or triaxial woven fabrics as they call them). It seems that such fabrics are more stable under diagonal strain than 2-way 2-fold fabrics, and so have been used in such practical applications as parachutes.

REFERENCES

1. B Astle, Pantactic squares, Math. Gazette 49(1965), 144-152; MR 32#1608.
2. Barber-Coleman Company, Textile machinery publicity literature, Rockford, Illinois 1979.
3. C J Bouwkamp, P Janssen and A Koene, Note on pantactic squares, Math. Gazette 54(1970), 348-351.
4. C R J Clapham, When a fabric hangs together, Bull. London Math. Soc. 12(1980), 161-164; MR 81g:51012.
5. T C Enns, Efficient algorithms determining when a fabric hangs together, Geom. Dedicata.
6. E Gand, Le Transpositeur ou l'Improvisateur de Tissus, Paris 1871.
7. Branko Grünbaum and G C Shephard, Satins and twills: an introduction to the geometry of fabrics, Math. Magazine 53(1980), 139-161; MR 82k:52017.
8. Branko Grünbaum and G C Shephard, A catalogue of isonemal fabrics, Proc. N.Y. Academy of Sciences (to appear).
9. Branko Grünbaum and G C Shephard, Isonemal fabrics of periods greater than eight (in preparation).
10. Branko Grübaum and G C Shephard, Isonemal fabrics, American Math. Monthly (to appear).
11. Branko Grünbaum and G C Shephard, Geometry of fabrics, Abstract 757-D1, Notices of the American Math. Soc. 25(1978), A-462.
12. J A Hoskins, Factoring binary matrices: a weaver's approach, Lecture Notes in Mathematics 952, Springer-Verlag, Berlin—New York, 1983, 300-326.
13. J A Hoskins and W D Hoskins, The solutions of certain matrix equations arising from the structural analysis of woven fabrics, Ars Combinatoria 11(1981), 51-59; MR 83h:05027.
14. J A Hoskins, W D Hoskins, A P Street and R G Stanton, Some elementary isonemal binary matrices, Ars Combinatoria 13(1982), 3-38.
15. J A Hoskins, C E Praeger and A P Street, Twills with bounded float length,

Proc. Adelaide Conference on Combinatorial Mathematics (to appear).

16. J A Hoskins, R G Stanton and A P Street, Enumerating the compound twillins, Congressus Numerantium (to appear).

17. W D Hoskins and A P Street, Twills on a given number of harnesses, J. Austral. Math. Soc. (Ser. A) 33(1982), 1-15; MR 83k: 05034.

18. W D Hoskins and R S D Thomas, Conditions for isonemal arrays on a Cartesian grid, Lin. Algebra and Appl. (to appear).

19. E Lucas, Application de l'Arithmétique à la Construction de l'Armure des Satins Réguliers, Paris 1867 .

20. E Lucas, Principii fondamentali della geometria dei tessuti, L'Ingegneria Civile e le Arti Industriali 6(1880), 140-111, 113-115.

21. E Lucas, Le principes fondamentaux de la géometrie des tissus, Compte Rendu de l'Association Francaise pour l'Avancement des Sciences 40(1911), 72-88.

22. O T Mason, Anyam Gila (Mad Weave), U.S.National Museum Proc. 36(1909), Washington D.C.

23. H Nisbet, Grammar of Textile Design, 3rd Ed., London 1927.

24. J J Pedersen, Some isonemal fabrics on polyhedral surfaces, The Geometric Vein, The Coxeter Festschrift (ed. C Davis, B Grünbaum and F A Sherk),New York 1981, 99-122.

25. E S Selmer, Doubly periodic arrays, Computers in Number Theory, Proceedings of the S.R.C. Atlas Symposium No. 2 (ed. A O L Atkin and B J Birch),Oxford 1969.

26. S A Shorter, The mathematical theory of sateen arrangement, Math. Gazette 10(1920), 92-97.

27. J E A Steggall, On the numbers of patterns that can be derived from certain elements, Messenger of Math. 37(1908), 56-61.

28. H J Woods, The geometrical basis of pattern design, Textile Institute of Manchester Journal 26(1935) (Section 2, Transactions): Part I - Point and line symmetry in simple figures and borders, T197-T210; Part II - Nets and sateens, T293-T308; Part III - Geometrical symmetry in plane patterns, T341-T357.

Branko Grünbaum,
Department of Mathematics,
University of Washington,
Seattle,
Washington 98195,
U.S.A.

G C Shephard,
School of Mathematics and Physics,
University of East Anglia,
University Plain,
NORWICH, NR4 7TJ.